BALLPARKING

PRACTICAL MATH
FOR
IMPRACTICAL SPORTS QUESTIONS

AARON SANTOS

RUNNING PRESS
PHILADELPHIA · LONDON

Published by Running Press,
A Member of the Perseus Books Group

Books published by Running Press are available at special discounts for bulk purchases in the United States by corporations, institutions, and other organizations. For more information, please contact the Special Markets Department at the Perseus Books Group, 2300 Chestnut Street, Suite 200, Philadelphia, PA 19103, or call (800) 810-4145, ext. 5000, or e-mail special.markets@perseusbooks.com.

ISBN 978-0-7624-4345-1
Library of Congress Control Number: 2012934948

E-book ISBN 978-0-7624-4501-1

9   8   7   6   5   4   3   2   1
Digit on the right indicates the number of this printing

Cover and Interior Illustrations by Mario Zucca
Cover and Interior Design by Joshua McDonnell
Typography: Avenir, Bembo and Helvetica

Running Press Book Publishers
2300 Chestnut Street
Philadelphia, PA 19103–4371

Visit us on the web!
www.runningpress.com

CONTENTS

I. INTRODUCTION - 5

II. TEE BALL ESTIMATIONS - - - - - - - - - - - - - - - - - - - 14

III. SCALING THE RUNGS OF LITTLE LEAGUE - - - - - - - - - - 48

IV. WELCOME TO VARSITY - - - - - - - - - - - - - - - - - - - 60

V. THAT'S HOW THE BALL BOUNCES - - - - - - - - - - - - - 163

VI. CONGRATULATIONS! YOU'VE MADE IT TO THE PROS - - - - 190

VII. HANGING UP YOUR SPIKES - - - - - - - - - - - - - - - - - 210

THANK YOUS - 211

ABOUT THE AUTHOR - 212

INDEX - 213

Introduction

Over the last ten years, sabermetrics has made unprecedented leaps in both the amount and quality of statistics used to analyze athletic performance. Once considered trivial and relegated to the back of baseball cards, statistics are now a vital tool for any general manager, with some teams investing millions of dollars to research better metrics by which to gauge potential prospects. As statisticians race to derive better metrics and, ultimately, generate billions in revenues for their teams, the talking heads of ESPN, sports blogs, and talk radio disseminate bits of statistical knowledge and mathematical terminology for even the casual sports fan to absorb, generating what some might call a more enlightened culture. Given this unprecedented epoch in sports history, the time has come for a book detailing the inner workings of this burgeoning field so that everyday laymen can share in the joys and mutual enlightenment mathematical rigor brings about.

This is not a book about sabermetrics. This is a book about playing a game, about a pleasurable pastime, about the humor in tomfoolery. In short, this is a book about math.

Roy Campanella once said, "You gotta be a man to play baseball for a living, but you gotta have a lot of little boy in you, too." Make all the pedophile jokes you want, but Mr. Campanella has a point. In mathematics, as in sports, those having the most fun—the ones who treat their endeavor as a game rather than a chore—are often the most successful. For many of us, sports remains a happy diversion well after we've taken our last gym class. Sadly, the same cannot be said for mathematics. I'm not sure what grade our teachers beat it out of us, but at some point mathematics became a chore. It wasn't always this way. Those same ancient Greeks who founded the Olympics also valued mathematics, happily developing number theory, mathematical analysis, conic sections, and Euclidian geometry, just to name a few. Again, make all the pedophile jokes you want, but the Greeks were on to something. There's lots of fun to be had playing with mathematics, and my goal in this book is to illustrate exactly how.

In the 2005 movie *Fever Pitch*, Drew Barrymore's character asks some students if they ever look at license plates, addresses, or telephone numbers and rearrange the numbers in their heads, to which one student replies, "Oh, my God, she knows my secret shame!" Even though bad teachers and math-phobic bullies may have beaten the geeky out of us long ago, playing with numbers and other mathematical concepts is something humans do naturally. Although our culture at large is profoundly innumerate, there may yet be hope for the jaded math geek in all of us. There's been a sharp rise in the popularity of geek humor through websites like xkcd, GraphJam, CollegeHumor, and ThinkGeek and even some graphical-function humor by mainstream comic Demetri Martin. Perhaps the biggest indication of a geek uprising is the increasing prevalence of sabermetrics stat geeks in the professional sports world, which, let's face it, hasn't exactly been known as an intellectual haven.

"But," you say, "I'm not much of a math person. Why is this book for me?" Good question, Reader. You don't need a 40-inch vertical to enjoy playing pickup basketball, and you sure don't need a PhD to enjoy playing with math-ematics. As a casual sports fan, you've probably pondered many of the questions asked in this book. Could Michael Phelps swim with cement loafers? Why don't hockey teams just shove a morbidly obese guy in the goal so no pucks can get in? How many STDs did Wilt Chamberlain catch after sleeping with 20,000 women? Okay, maybe you didn't ask these exact ques-tions, but I'm willing to bet you've asked similar ones, and that's what this book is about: teaching you to answer your own sports questions by going through some hopefully entertaining examples. Before we delve into prob-lems, I should give you a little bit of background on some of the mathematical tools I'll use throughout the book. Following is a brief interlude on scientific notation, units, and other goodies.

Scientific Notation

A few years back, an Internet meme compared the money made by Alex Rodriguez to that of Bill Gates. As it turns out, A-Rod would have to play at his current salary for 2,000 years in order to be as rich as Gates. This result is very easy to derive using math we learned in third grade. Why, then, is the result so surprising?

Most numbers in sports are fairly palpable. DiMaggio's 56-game hitting streak, Wilt's 100 points in a game, and Ted's .406 batting average are all numbers we experience in everyday life. Other numbers, like A-Rod's $250 million salary, are more difficult to conceptualize because most of us don't commonly use numbers this large. To deal with very large—and very small—numbers, scientists and engineers use scientific notation. The idea is fairly simple. The number 250,000,000 has a lot of zeros in it, and we could save space and time by not writing them all. Mathematically, this number is equivalent to $2.5 \times 10 \times 10 \times 10 \times 10 \times 10 \times 10 \times 10 \times 10$. So we don't have to write out all the 10s, mathematicians invented exponents as shorthand. Using exponents, you can write $10 \times 10 \times 10 \times 10 \times 10 \times 10 \times 10 \times 10$ succinctly as 10^8. Similarly, we can write the number of dollars in A-Rod's contract as:

$$250,000,000 = 2.5 \times 10^8$$

This is known as scientific notation. The standard form has some decimal number between 1 and 10 multiplied by 10 to some power. The power tells you how many spaces to move the decimal point. For example, 10^7 means move the decimal point 7 places to the right, whereas negative powers like 10^{-8} mean move the decimal point 8 places to the left.

Scientific notation is very useful for comparing large numbers. The exponent, sometimes called the "order of magnitude," quickly gives a rough idea of the size scale. For example, by themselves, 37,000,000,000,000,000,000,000,000,000,000,000 and 5,100,000,000,000,000,000,000,000,000,000 just look like big numbers. In scientific notation these numbers are written compactly as 3.7×10^{34} and

5.1×10^{30}, respectively. Now, you can quickly see that the exponents differ by 34−30 = 4, which means the first is about 10^4 = 10,000 times larger than the second. In some cases, scientific notation is not just useful but absolutely essential. A very large number like 3.2×10^{5,000,000} would take over 1,000 pages to write out by hand! From this point on, I'll be writing all large numbers in the book using scientific notation.

Even when working with familiar numbers, it helps to practice formally estimating things. If you're going to Spain to participate in the running of the bulls, it's probably a good idea to know how much of a head start to take. For problems like this, having a firm mathematical background will only increase the accuracy of your estimate and, in this example, may help you live longer. It is in this regard that scientific notation possesses a real advantage: it's great for expressing real-world numbers.

Numbers measured in the real world are, to borrow a term from a former president, "fuzzy." [1] Consider the Tour de France, the grueling three-week bicycle trek through 3,600 kilometers (~2,000 miles) of French countryside. Do the riders travel exactly 3,600 kilometers? Almost certainly not! Even if they never veer off course for a pee break, they're constantly swerving to avoid and pass other riders, making their path less straight and somewhat longer. So why don't we say the actual distance is 3,603.1415926 kilometers? Even if we wanted to write down all these extra numbers, measuring the distance traveled with that kind of precision is impossible. In this case the extra digits have no meaning. Some riders might travel a few inches over 3,600 kilometers. Others might travel tens of kilometers more. It's impossible to define what these last few digits are because they'll change for each rider in the race, but the 3,600-kilometer figure implies that those last two digits are exact. To truly understand how fuzzy the number is, we'd need to attach an error range. [2] For example, we could say the actual distance traveled is 3,600 kilometers, plus or minus 100 kilometers, but there's a more compact way.

---

[1] I suspect NFL draft prospects take advantage of the fuzziness inherent in real-world numbers. Although I doubt draft prospects would outright lie about their Combine scores to a scout, it's very probable that they only report their best results. I'm guilty of this myself. Despite the fact that I'm consistently measured at 5 feet, 11 inches tall, I always tell people I'm 6 foot because this is the tallest I've ever been measured.

Scientific notation takes care of the problem for us. Consider the numbers 3.6×10^3 kilometers and 3.600000000×10^3 kilometers. Mathematically, these two numbers are identical, but they should be interpreted quite differently. Each of the digits listed is assumed to be significant, meaning that they've been measured or at least estimated to good precision.[3] At most the last digit might be a little off. With this interpretation, one can readily see that the first number implies an error range of 100 kilometers, whereas the second implies the measurement is accurate to within 1 millimeter.

Units

There are many ways to provide context for numbers. For example, I could say that A-Rod's $250 million contract is worth 68 times his weight in gold. I could just as easily state that A-Rod's contract money could cover the entire playing field at Yankee stadium with pennies over 600 times. I could equivalently say the same money converted into dollar bills would fill ten Olympic-sized swimming pools. I could keep going with different examples, but fundamentally these are just different units. "Dollar bill-filled Olympic swimming pools" is really just another unit of money like yen or drachmas. The only reason we give preference to certain types of units like meters and kilograms is that people are familiar with these.[4] These units provide a standardized context.

---

2 I'm a big proponent of using error ranges in everyday life. If you've ever opened a bag of chips that was more air than chip, you probably understand why. Like all real-world numbers, those listed on the side of food packages are not exact. In general, the actual number of calories in a piece of food deviates from the mean value listed on the package. If you're on one of those carb-counting diets, you probably want to know whether or not that baked potato has 100 calories more than what's listed on the packaging. The same can be said for the "calories burned" meter you see on some exercise bikes. Manufactured and agricultural goods always exhibit fluctuations. Including the error range on the packaging would lead to more truth in advertising, and, I would argue, a slightly more mathematically enlightened populace.

3 For aesthetic reasons, I generally like to keep two significant figures when I use numbers in scientific notation. Many of the problems are only order of magnitude estimates. For these problems, know that the second figure is largely superfluous.

4 I prefer to use the International System of Units (i.e., kilograms, meters, etc.) because it makes unit conversions much easier. Unfortunately, sports often use English units (i.e., pounds, feet, Fahrenheit, etc.), which are a wretch to convert. Because it seems unnatural to say the distance between bases is 27.43 meters or that football offenses need to travel 9.14 meters for a first down, I switch back and forth between unit systems depending on the situation. For readers who are only familiar with one system of units, I list both values in cases where confusion may arise.

You might wonder why we have so many different units. For example, why do we measure distance in inches, centimeters, nanometers, miles, kilometers, parsecs, and light-years? Why can't we just pick one unit and stick with it? The reason is partly so we can work with familiar numbers. Saying that a pitcher's mound is 1.9×10^{-15} light-years away from home plate would be mathematically correct, but this number isn't nearly as easy to grasp as saying 60.5 feet. Again, it's an issue of context because one unit provides more perspective than the other.

Whether we like it or not, we're stuck using different units, so we'll have to learn how to convert from one form to another. Most of us learned how to do this in high school chemistry class. Let's say you're converting yards to miles. There are 1,760 yards in 1 mile. To convert units you simply multiply the "yards" times "miles per yard." When you do this, the "yards" will cancel, leaving units of miles. We can use this to shed new light on old numbers. Take Emmitt Smith's NFL record 18,355 career rushing yards. Using the conversion factor above, we can calculate the distance Emmitt ran in miles:

$$\left(18,355 \text{ yds}\right) \times \frac{\left(1 \text{ mi}\right)}{\left(1760 \text{ yds}\right)} = 10.4 \text{ mi}$$

This means Emmitt Smith collectively ran a little less than a half-marathon over his NFL career. We can do the same calculation for Jerry Rice's 22,895 career receiving yards (~0.5 marathons) and Dan Marino's 61,361 career passing yards (~1.3 marathons).

Tools

Unit conversion is not difficult, but it can be annoying if you do it often enough. Fortunately, we live in an age when many people have web access literally at their fingertips in the form of smart phones, laptops, and so forth, and there are plenty of websites that will convert units for you. If you're planning to make lots of estimates, I highly recommend using some online unit converter like those found on Google and WolframAlpha. By simply typing your expression into the input box, you instantly obtain the numerical result given in international units. Whether you enter your numbers in feet, meters, or light-years doesn't matter; the engine will do the unit conversion for you and still calculate the correct answer. If you'd prefer your answer in a different unit, all you have to do is type "in [insert preferred unit]" after the expression and the output will be expressed in your desired unit. As best I can tell, the only drawback of converters like these is that they can't work with some of the more esoteric units in the book like "teeth per hockey player."

For many calculations, it helps to look up numerical data like the speed of light or the radius of the Earth. Despite the complaints of many high school teachers, Wikipedia is very reliable for this purpose. Another personal favorite is The Physics Factbook on hypertextbook.com, which features everything from the density of milk to the force of a windmill slam dunk with references or calculations that you can use to fact check.

Perhaps even more important than listing websites that are useful is listing the websites that are downright awful. Avoid Yahoo Answers, Ask, Answers, and any other site where users can submit answers without any verification. These sites are notoriously unreliable and should be avoided like Albert Haynesworth on an eating binge.

Warning

I will, on occasion, include problems that require the use of various equations and notations (integrals, derivatives, summations, etc.) typically found in advanced math and physics courses. I include these problems because I find them particularly fun and enlightening, and it would be a shame to let a little bit of scary math deprive you from seeing them. However, when these problems pop up, I'll do my best to give you a qualitative feel for the underlying math while simultaneously shielding your eyes from the messy particulars, even if only to ensure my editors that the casual reader flipping through this text while passing through a bookstore will not be put off by the messy notation and run away screaming.

Disclaimer

Physics is a complicated beast, and writing a fun layperson book (with math, no less) is necessarily going to neglect some of the complications. I'll try to mention it every time I do something that's a little questionable, but that's what we, as physicists, do. We make a lot of assumptions. Very few of the physics problems worth doing are mathematically tractable, so we look for approximations and then make preliminary calculations to see if these approximations make sense. If they do, "Yay! We've just completed the first step to solving our problem!" More often than we'd like, the preliminary calculations show that our assumptions are faulty and we have to start over. In some sense this book is all about the preliminary calculations. Some are insightful and others are downright silly, but they should give you a feel for the types of mental games you can play with math and a little bit of physics knowledge. Just remember that what I'm showing here is only the first step in a process. I don't pretend my answers are necessarily correct, and neither should you. My hope is that you'll think about the problems yourself and derive your own answers. If you can come up with better estimates than what I've done here, I'll feel that I've done my job.

TEE BALL ESTIMATIONS

As human beings, we constantly make mental models of the world around us. When you pick up a golf club, you generally don't imagine the structure of the biomolecules in the wood of the club head, what type of tree the wood came from, or how the rivers, streams, and rainfall in the environment surrounding the tree contributed to its elastic properties, even though these factors ostensibly affect the distance your shot will travel. To comprehend the world, we have to simplify it: bend knees, keep head down, swing club, ball go. Our brains are innately wired to eliminate information deemed irrelevant. To do so, we make simplifying assumptions on the fly, often without being aware we're doing it. (If you doubt this last point, try flipping ahead to the problem titled "Rodeo Clown.") Whether we want to or not, we're always making assumptions about the world and interpreting our senses through a simplifying filter, so we shouldn't be ashamed of consciously using approximate numbers to calculate estimates even though we know the numbers aren't exactly correct. By making simple estimations, you gain a better understanding of the vastly complicated real world and, in the process, familiarize yourself with the role mathematics plays in governing the happenings of everyday life.

To make good estimates, it helps to follow these simple rules:[5]

1. Start with what you know. Let's say you're calculating the number of teeth hockey players in the NHL lose each year. You probably don't know the total number of players, so you can't start there. If you follow the sport, you probably have a rough idea of the number of players per team and the number of teams in the league. Even if you don't, you can always look this up.[6] Starting with simple, known facts, you can derive other facts.

---

5 This estimation technique was popularized by physicist Enrico Fermi and is often called "the Fermi method."

6 There's no shame in looking up numbers. There are many quantities that are a pain to calculate and can easily be found via a quick web search. Unless instructive, I'm not going to calculate anything you can find online in under two minutes.

2. Build a path by canceling units. Let's say you've estimated one tooth lost per player, 30 players per team, and 30 teams in the league. With this data, you can build a path to the answer by canceling units:

$$\frac{1 \text{ tooth lost}}{1 \text{ player}} \times \frac{30 \text{ players}}{1 \text{ team}} \times 30 \text{ teams} = 900 \text{ teeth lost}$$

3. Use upper and lower bounds. If you're ever uneasy about one of your assumed numbers, you can always test its accuracy by setting upper and lower bounds. For example, let's say you didn't trust my one-tooth-lost-per-player assumption. You'd probably agree that the actual number lies somewhere between ten teeth per player, which would leave no teeth left after three years, and one tooth every ten years. Given these bounds, our estimate should be correct to within an order of magnitude (i.e., within a power of ten). Throughout the book I will set bounds like these whenever I feel an assumed number is questionable.

4. Be honest. We all have biases, and it's important not to let these affect our thought process. Never try to make a number come out the way you want it to. (I've been guilty of this on more than one occasion.) When in doubt, always use conservative estimates and put bounds on the answer.

5. Enjoy. Remember, this is just a game you're playing. If you never treat it as more than that, you'll be sure to come up with some fun and interesting facts.

Those Screwy Celtics

In a 1991 episode of the sitcom *Cheers*, Boston Celtics forward Kevin McHale grows obsessed with knowing how many bolts are in the Boston Garden basketball court, and the obsession ruins his game. If only Kevin had a copy of *Ballparking*. **How many bolts were in the basketball court at the old Garden?**

An NBA basketball court is 94 feet by 50 feet. If you check out the floor during a game, you'll notice there are about 10 floorboards running across the width of the court and 18 running down its length, giving a total of 180 floorboards. If each floorboard is held down by 4 bolts (one at each corner), there will be 720 total bolts.

A quick Google search shows there are actually 988 bolts. Presumably, my estimate is lower because I only counted the playing surface and not the several meters of floorboards that make up the sidelines.

Buck'n Bench

I wish I knew which was longer: ⅜, ½, ⅝. . . . You played in the NFL, Troy: what's longer, ⅝ or ½?

—**Fox Sports announcer Joe Buck discussing the size of cleats during an NFL game**

People in the sports world have a reputation for being dumb jocks. With comments like the one above, it's not hard to see why. Happily, there's at least one athlete who can estimate with the best of them. In an interview, Reds Hall of Fame catcher Johnny Bench estimated he'd squatted 500,000 times throughout his career. **Was Mr. Bench's estimate close?**

Bench played almost 20 years with the Cincinnati Reds, not counting minor league and nonprofessional games. He played about 150 games per year. In a typical game, the ball is pitched about 100 times, meaning Bench squatted about 100 times per game. His total number of squats is likely to be about double this since he's also catching for practices. From these numbers, we can estimate how many times he squatted over his career:

499,376....
499,377....
499,378...

# of squats = 2 · (# of squats per game) · (# of games per year) · (# of years)
= 2 · (100 squats per game) · (150 games per year) · (20 years)
= 600,000 squats.

Well done, Mr. Bench!

Play Hockey, Save on Toothpaste

A friend was sitting by the Washington Capital's bench when one of their players got hit in the mouth with a puck. In his words, "It was disgusting. When he came back to the bench and took his mouth guard out, his entire face slid down." Needless to say, the unfortunate player was missing a few teeth. Given the stereotypical grinning, gap-toothed hockey mouth, one might say it's just an occupational hazard. Even though the game has evolved, tooth loss has remained fairly constant. This might be somewhat surprising because helmets weren't always around, but then again, neither were composite sticks, steroids, or Zdeno Chara fights. Many players take the omission of a few incisors as a badge of honor, but with this warrior mentality comes a surprising financial benefit. **Since its inception, how much money has the NHL and its players saved on toothpaste?**

There are 32 teeth in a human mouth. We can subtract 4 to get 28 teeth because many people have their wisdom teeth removed. If we only count original teeth (i.e., no fake implants), then the most teeth a hockey player can lose is 28. This is only an upper bound because even hockey veterans have some of their original teeth. Some seasons a player gets lucky and loses none, whereas other seasons he might lose 10 on one slap shot. I'll assume an average of 1 tooth lost per player per season, as this fits nicely with our upper bound:

Upper Bound: 28 teeth
Assumed Average: 1 tooth per player per season
Lower Bound: 0.1 teeth

The NHL started in 1917 with 4 teams, but expansion and contraction caused the number of teams to fluctuate over the years.[7] Since the 2000–2001 season there have been 30 teams, so we might estimate that there have been an average of 15 teams per season over the last 90 years. Teams dress 18 skaters and 2 goalies to give a total of 20 players per team. At 28 teeth per player, a maximum of 560 teeth can be lost per team per season, more if you consider players brought up from the minors. Using the average, it's much more likely that about 20 teeth are lost per team each year. From this, one can estimate the total number of teeth lost in NHL history:

# of teeth = (# of teeth lost per team) · (# of teams per year) · (# of years) = (20 teeth per team per year) · (15 teams) · (90 years) = 27,000 teeth.

Assuming tooth dimensions of about 1 centimeter by 7 millimeters by 5 millimeters, you find almost 10 liters (~2.5 gallons) of teeth have been lost by NHL players over the years. Since there are 28 teeth per mouth, roughly 1,000 mouths worth of teeth have been lost. A player might live another 50 years not having to brush those missing teeth. Because toothpaste costs $3 per tube and tubes last about 1 month, we can estimate that NHL players have collectively saved

(1000 mouths) · ($3 / mouth · month) · (12 month / years) · (50 years) = $1,800,000.

That's $1.8 million dollars NHL players saved on toothpaste, although this figure must pale in comparison to the collective amount they've spent on other dental care.

---

7 The famed "Original Six" didn't show up until 1942.

Is This Seat Taken?

In the 1985 movie *Brewster's Millions*, Richard Pryor plays Monty Brewster, a career minor league baseball pitcher who will inherit $300 million provided he can spend $30 million within 30 days. To spend part of the money, Brewster hires the New York Yankees to play an exhibition game against his own team, the Hackensack Bulls. In real life, you'd probably need some pretty good connections to set up an exhibition game with a Major League team. If it didn't work out, Brewster could instead have spent the money on my childhood fantasy: buying every ticket to a game. Think of it. It'd be just you and the players. Hanging out. Chatting. Getting easy autographs. You'd get premium service from the vendors. There'd be absolutely no one else in the stands. Come to think of it, it'd be a lot like going to a Pirates game. **What's a quicker way to spend money: paying the Yankees for an exhibition game or buying all the tickets in Yankees Stadium?**

The average price for a seat at the new Yankee Stadium is about $75, and the stadium can hold about 52,000 people. From this, it's easy to see that the total cost of buying every seat would be

$$(\$75 \text{ / ticket}) \cdot (52{,}000 \text{ tickets}) = \$3{,}900{,}000.$$

The Yankees 2011 payroll is roughly $197 million, and there are 81 home games. Assuming that hiring the team for an exhibition game would cost as much as a regular season game, we can calculate the cost per game:

$$(\$197{,}000{,}000) \text{ / } (81 \text{ games}) = \$2{,}400{,}000 \text{ per game.}$$

Brewster can spend 50 percent more by buying every ticket to the game.[8] In fact, he could have spent almost all of the money by buying every seat on a long home stand.

---

8 Some might wonder whether this answer should be obvious given that the Yankees have to take in more money than it costs to pay the players to be financially viable, but this neglects the large amount of money teams make with advertising revenue, television contracts, and merchandise.

Ratatouille Franks

Ahh, the ballpark frank! The quintessential baseball food. There's nothing like spending a hot summer day watching a ballgame with a hot dog in one hand and a cold beer in the other. Unless, of course, you remember all those elementary school rumors about hot dogs containing up to 5 percent rat droppings. **If this rumor were true, what's the total mass of rat poo consumed in MLB ballparks each season?**

Hot dogs typically weigh about 100 grams. Not everyone who goes to a baseball game eats a hot dog, but the actual percentage of hot dog eaters is almost certainly between 1 percent and 100 percent of people. Using these bounds, we can assume about 10 percent of people at baseball games eat a hot dog or other sausage-like entity.

Upper Bound: 100%
Assumed Average: 10%
Lower Bound: 1%

Baseball stadiums can fit anywhere between 37,000 (the Oakland Coliseum) and 56,000 (Dodger Stadium) fans. I'll assume an average of 40,000 fans attend each game. There are 81 home games per team each year and 30 teams in the majors. From this, we can estimate the total mass of rat droppings eaten each year:

$$(5\%) \cdot (100 \text{ g / hot dog}) \cdot (1 \text{ hot dog / 10 people}) \cdot$$
$$(40,000 \text{ people / game}) \cdot (81 \text{ games / team}) \cdot (30 \text{ teams})$$
$$= 49,000 \text{ kg } (\sim 54 \text{ tons}).$$

If true, this would mean fans consume 54 tons of rat droppings each year at Major League Baseball stadiums.

The Crying Fields of Dreams

Tom Hanks, speaking as Manager Jimmy Dugan in *A League of Their Own*, famously said, "There's no crying in baseball." I suspect Mr. Dugan never saw *Field of Dreams*. According to a 2005 *Entertainment Weekly* reader poll, *Field of Dreams* was the fourth highest-ranking "guy-cry" film.[9] If you've ever played a game of catch with your dad—and even more so if you've *never* played a game of catch with your dad—you know why. Fight it though you may, this movie is bound to get the waterworks flowing. **How many male tears have been shed for *Field of Dreams*?**

9 Ironically, the usual trick of thinking about baseball to avoid dealing with emotionally complex situations only makes matters worse here.

I'm going to assume the vast majority of people who have seen *Field of Dreams* are American. That gives an audience of 150 million American males. If you asked 100 of your male friends, I'm willing to bet at least 1 of them will admit seeing *Field of Dreams*. We can take this as our lower bound and again assume 10 percent as the average.

Upper Bound: 100%
Assumed Average: 10%
Lower Bound: 1%

I'll assume the average viewer has seen the movie twice. If only to avoid the inevitable tear-fest, most men will probably not watch it more than a handful of times. The average number of viewings is almost certainly less than 20, so our estimate of twice per viewer seems reasonable. Even if you're trying to hold the tears in, I suspect you'll have at least five tears leak out as soon as Ray Kinsella utters the words, "Hey . . . Dad? You wanna have a catch?" From this we can estimate the number of tears:

$$(5 \text{ tears / movie}) \cdot (2 \text{ movies / viewer}) \cdot (1 \text{ viewer / 10 men})$$
$$\cdot \, (1.5 \times 10^8 \text{ men}) = 1.5 \times 10^8 \text{ tears.}$$

This means American men have shed about 150 million tears watching Kevin Costner. That's enough teardrops to fill 15 beer kegs.

Long Time, First Time

Big O: We're going to Dave from Medford.

Caller: Hi, Big O. Long-time listener, first-time caller.

Big O: What's your question, Dave?

Caller: I think we should trade John Lackey for Roy Halladay.

The above dialog is an amalgamation of calls I've heard on WEEI's *The Big Show*. Frankly, I don't know how sports talk-radio hosts stay sane. Whether it's complaints about how the refs *only* go against your team, whining about how the manager who just won you two championships should be fired for leaving the starter in one batter too long, or suggestions that the front office make a trade so unbalanced that even an EA Sports video game would reject it, the sports-talk host always has to listen and—hopefully—politely explain why the caller is in the wrong. The above example is particularly striking because the caller seems to lack any sort of theory of mind in regard to the general managers of other teams.[10] How does a host politely explain the patently obvious fact that no person in their right mind would trade Roy Halladay straight up for John Lackey?

This may seem like it has very little to do with estimation, but the simple fact is that estimation has as much to do with basic logic as it does with numbers, so I'd be remiss if I didn't at least discuss some of the logical inconsistencies that come up during a typical sports-talk phone call. Consider the logic behind one caller's comparison between Lance Armstrong's seventh Tour de France win and speed skater Eric Heiden's five gold medals in the 1980 Lake Place Olympics:

---

10 Wikipedia defines "theory of mind" as "the ability to attribute mental states—beliefs, intents, desires, pretending, knowledge, etc.—to oneself and others and to understand that others have beliefs, desires, and intentions that are different from one's own." Humans typically develop theory of mind before age five.

Caller: Here's why Lance is way better than Heiden. In 1986 Heiden raced in the Tour de France and didn't finish . . .

Although the caller is certainly knowledgeable about Heiden's cycling career—Heiden crashed five days before the finish—the argument itself doesn't follow basic logic. An athlete shouldn't be judged solely by his performance in a competition that's not even his primary sport. That would be like saying Michael Jordan is less of an athlete than Mario Mendoza because he only made it to Double-A baseball. Furthermore, why isn't the same logic applied to Lance Armstrong? Eric Heiden was inducted into the US Bicycling Hall of Fame, but I've yet to see Lance Armstrong strap on a pair of skates. This is not to denigrate Armstrong's accomplishments, of which there are many. Which of the two men is a better athlete is certainly up for debate, but what's not debatable is that the caller's logic is full of holes.[11]

To be fair, I make my fair share of illogical statements when it comes to sports, so perhaps I shouldn't be so quick to criticize sports-talk callers. Then again, I don't go out of my way to spread my ignorance over the radio, so I think I'm good on this one.[12] Although I could go on a long diatribe about the logical fallacies sports-talk callers make (and often sports-talk radio hosts), I should probably get back to actual numbers. **How many illogical calls are made to sports radio stations each year?**

By perusing the Wikipedia page for "Sport Radio," you'll see there are about 100 sports radio stations. Not all of the shows allow you to call and state your opinions, so I'll assume there are eight hours worth of live call-in shows each day. Being somewhat conservative, I'd say there are about 2 illogical callers per hour because there are certainly fewer than 20 per hour and more than 2 every ten hours.

---

11 We'll discuss more on comparing athletes from different sports in "A 'Better' Stat."
12 I find spreading my ignorance through book sales to be much more profitable.

Upper Bound: 20 per hour
Assumed Average: 2 per hour
Lower Bound: 0.2 per hour

From this we can estimate the number of illogical calls made each year to sports radio stations:

$$(100 \text{ stations}) \cdot (2 \text{ calls / hour station}) \cdot (8 \text{ hours / day})$$
$$\cdot (365 \text{ days / year}) = 600{,}000 \text{ illogical calls per year.}$$

That's 600,000 illogical calls each year—and a whole lot of aspirin for the show hosts.

Mighty Buddha at the Bat

And lo God said, "Geno Auriemma, come forth!" But UConn came in fifth and God was pissed.

There's a time and a place for religion. Even my cynical heart is touched when athletes from rival teams kneel and pray for a seriously injured team-mate, but I get annoyed when I see basketball players blessing themselves before free throws or baseball players pointing to the heavens in thanks after a home run. Really? There's poverty, famine, wars, and genocide—none of which God has any desire to intervene on—but for some reason he's decided to cart his holy ass down from heaven and reel your would-be popup into the right-field bleachers with his magic golden fishing rod? Forgive me if I'm a bit skeptical, but if praying to God really helps an athlete's wishes come true, I'm pretty sure I would have hit some home runs in Little League. The most irri-tating part is that underneath the false piety lies the arrogant assumption that "God wants me to win more than that other guy." It's a lot like when siblings ask a parent to pick who s/he loves more, only in this case, the form of the question necessarily dictates that an answer will be given. Putting religion aside, there's still a question to be asked: **How many prayers are said in an attempt to improve athletic performance in each year?**

To narrow the question a little bit, I'm only going to consider the four major American team sports. Because professional athletes are largely a reflection of society, it's likely that around 90 percent are religious in some form or other. There are about 800 MLB players, 1,600 NFL players, 450 NBA players, and 600 NHL players, each of whom plays 162, 16, 82, and 82 games per season, respectively. This gives a total of about 10^5 games athletes play each year.

Even with the 90 percent religious figure, I suspect a significantly smaller fraction of players actually pray to help their performance on any given day, but the actual number likely falls between 1 and 100 percent, so I'll assume an average of 10 percent.

Upper bound: 100%
Assumed Average: 10%
Lower bound: 1%

With this assumption, we can easily compute the number of prayers said each year:

$$(10^5 \text{ games played}) \cdot (1 \text{ prayer} / 10 \text{ games played})$$
$$= 10,000 \text{ prayers.}$$

That's 10,000 prayers each year asking God to pick sides.

Pippen Tippin'

A web search for the phrase "celebrity tippers" will bring up a list of celebrities, including some professional athletes, and a description of how they compare in the service-tipping department. On the good-tipper list are such notables as Marcus Allen, Andre Agassi, Charles Barkley, Joe Frazier, Steffi Graf, Peyton Manning, John Stockton, and Warren Sapp, while the bad-tipper list includes Derrick Coleman, Buster Douglas, Allen Iverson, John Randle, and Scottie Pippen. From a purely phonological standpoint, I love the thought of being able to say "Stiffin' Pippen" whenever referring to the former Bulls forward, but it's unclear the nickname is deserved. There's no easy way to confirm the accuracy of the list, so it's entirely possible there's a Scottie Pippen doppelganger out there somewhere ruthlessly stiffing the wait staff. Still, there's something about having legions of adoring fans that might make an athlete feel he's above common courtesy. Once you reach a certain level of fame, tipping might start to seem somewhat superfluous. **How much would an athlete have to spend on a meal to have a good tip be worth more than a simple autograph?**

The question is somewhat ill defined. The value of a celebrity's autograph depends on many factors: the degree of famousness, rarity, whether or not the person is still living, etc. Thumbing through a copy of *SkyMall*, you can find Michael Jordan's autographed jersey for $1,999.99 and a boxing glove signed by Muhammad Ali for $2,499.99. These prices may seem exorbitant, but you have to remember we live in a world where someone once bid $10,000 for Luis Gonzalez's discarded bubble gum. To be fair, most athletes' signatures are worth appreciably less. Professional athlete autographs generally run between $50 and $1,000. Assuming a 20 percent gratuity—considered good by most wait staff I know—a celebrity athlete would have to eat anywhere from $250 to

$5,000 worth of food for the tip to be worth more than the John Hancock. Even if you're Joey Chestnut, that's an awful lot of money to spend on dinner. So the next time a celebrity athlete stiffs you on the bill, remember this: these millionaires are not giant douche-y cheapskates but rather simple people who believe their discarded baby back ribs serve as sufficient payment. Sometimes the best tip can be simply signing the credit card statement rather than paying in cash.

Just Retire the Whole Team Already

Given how profitable Major League Baseball is, it's tempting to believe it will be around forever. Sadly, that's not necessarily the case. Professional sports history is littered with corpses of now-defunct leagues:

- **The National Association of Professional Base Ball Players (1871–1875)**
- **American Basketball Association (1967–1976)**
- **International Hockey League (1945–2001)**
- **United States Football League (1983–1985)**
- **XFL (2001)**

This list is by no means exhaustive. Even if MLB lasts well into the next millennium, there's still a problem it will face: retired numbers. Retired numbers are the global warming of Major League Baseball. We know it's going to be a problem, but no one seems to care or want to do anything about it. **How long will it be until a MLB team runs out of one- and two-digit numbers?**

Admittedly, numbers don't stop after 99, but if MLB lasts long enough, it will face a philosophical crisis. Will they go to three-digit numbers? Negatives? Decimals? Fractions? Regardless of what the league decides, the first team to experience the problem will most likely be the Yankees. As with most statistical categories, the Yankees lead the league in retired numbers. At present, 16 numbers have been taken off the board.[13] Over the next ten years Derek Jeter's #2 and Joe Torre's #6 will likely be retired, leaving the team with no single-digit numbers. That'll give 18 retired numbers in about 100 years, or roughly 0.18 numbers per year.

Active MLB rosters hold 25 players for most of the season but expand to 40 in September. This means a team with 61 retired numbers will not have enough for every player on the roster.[14] As such, the Yankees can only retire 45 more numbers before running out. Dividing this by the rate of numbers retired, we get the time it will take for them to run out of numbers:

$$(45 \text{ numbers}) / (0.18 \text{ numbers} / \text{year}) = 250 \text{ years.}$$

In roughly 250 years, the Yankees will run out of numbers to retire. At that time Derek Jeter will be the same age George Washington is now. Fortunately, we'll all be dead by then, so we don't have to worry about it.

---

13 #1 Billy Martin, #3 Babe Ruth, #4 Lou Gehrig, #5 Joe DiMaggio, #7 Mickey Mantle, #8 Bill Dickey, #8 Yogi Berra, #9 Roger Maris, #10 Phil Rizzuto, #15 Thurman Munson, #16 Whitey Ford, #23 Don Mattingly, #32 Elston Howard, #37 Casey Stengel, #44 Reggie Jackson, #49 Ron Guidry, and #42 Jackie Robinson. All of MLB retired Jackie Robinson's #42.

14 I'm including all numbers between 0 and 99. The number #00 was famously worn by former Boston Celtic Robert Parish, but this number is rarely used in sports.

Skee Ball Game Theory

As a kid, I always wanted to win the stereo at the Dream Machine arcade in the North Dartmouth Mall. The stereo was worth 10,000 tickets.[15] **How long would it take to win that many tickets, and how much would it cost?**

Anyone who has spent time in an arcade knows the best way to get tickets is through skee ball. In addition to giving out lots of tickets, you can steal—er— acquire even more by pulling the tickets out of the slot very quickly. Usually one skee ball game would net you about 5 tickets. Back in the day, It cost 25 cents to play, and you might get a $2 allowance each week from Mom. At this rate, it would take

$$(10,000 \text{ tickets}) / [(5 \text{ tickets per } 25¢) \cdot (\$2 \text{ per week})]$$
$$= 5 \text{ years},$$

at a total cost of

$$(10,000 \text{ tickets}) / (5 \text{ tickets per } 25¢)$$
$$= \$500.$$

Given that the stereo was worth at most $100, I was probably better off spending the money on baseball cards.

---

15 I was about ten at the time, so my memory could be flawed on the exact number.

Eureka!

Only an idiot fills the tub all the way before stepping inside. As Archimedes famously noted, an object immersed in a fluid displaces the fluid upward (i.e., when the object goes down, the water goes up). But why doesn't a pool over-flow when a bunch of swimmers dive in it? **How many swimmers can fit in an Olympic-sized pool before it overflows?**

Let's imagine Michael Phelps tucked up into a ball. His radius would prob-ably be about 11 inches.[16] Using the formula for the volume of a sphere, Michael Phelps's total volume is about[17]

$$V = \frac{4\pi R^3}{3} \approx \frac{4 \cdot (3.14) \cdot (11 \text{ inch})^3}{3} \approx 0.09 \text{ m}^3$$

An Olympic-sized swimming pool has dimensions 50 by 25 meters, giving it an area of 1,250 square meters. When Phelps jumps in, the water level rises because some volume of water is displaced upward. We can figure out the height the water rises by dividing the volume displaced by the area of the pool:

$$\frac{V}{A} = \frac{(0.09 \text{ m}^3)}{(1250 \text{ m}^2)} \approx 7.0 \times 10^{-5} \text{ m}$$

---

16 If the 11-inch radius seems small, remember you need to imagine Michael Phelps curled into a ball that has no empty space. If you don't buy my logic, you can always calculate Phelps's volume using the fact that human bodies are about as dense as water, which has a density of 1,000 kg/m³. Dividing Phelps's mass of 88 kilograms (~165 pounds) by this density, one gets the same volume V = 0.09 cubic meters.

17 To a good approximation, a human body's volume is independent of whatever shape it happens to be in.

That's only 70 microns for every Phelps-sized swimmer who jumps in the pool.[18] There are about six inches between the surface of the water and the top of the pool. This means you'd need

$$(6 \text{ in}) / (7.0 \times 10^{-5} \text{ m per swimmer}) = 2{,}000 \text{ swimmers.}$$

Surprisingly, you'd need at least 2,000 swimmers to dive in before the pool overflowed. Evidently, our bodies are a lot smaller than you may have thought.

---

18 One micron is 1,000 times smaller than a millimeter.

Dog Hunt

I don't generally support violence against animals, but there is one major exception: the video game *Duck Hunt*. No, I'm not talking about shooting the ducks—I'm talking about shooting the dog.[19] I personally have shot that cocky little bastard at least 100 times. It's a natural human response given how he just stands there, laughing at you, mocking your plight as you sit hopelessly out of bullets while a terrified water fowl flies fleetingly into the distance. The only way that game would have been remotely bearable is if the dog actually responded when you shot him. **How many times have people shot the annoying dog in *Duck Hunt*?**

19 I'll let you insert your own Michael Vick jokes here.

I don't know anyone who bought *Duck Hunt* as a stand-alone game, so I'm going to assume most people who own it got it with the Nintendo Entertainment System (NES). According to Wikipedia, there were about 60 million NES units sold. Even if you include the times you go back to play for nostalgia's sake, you likely get at most 20 hours of game play out of *Duck Hunt*.[20] Taking this as an upper bound, I'll assume 3 hours of game play on average. Each round takes about one minute, and there are ten ducks per round. On average, you might miss two ducks per round, and, after each miss, Marmaduke's evil twin busts a gut for about three seconds. In that time you can usually get off about four shots. From these assumptions, we can estimate the total number of times the dog has been shot:

$$\text{(60 million people)} \cdot \text{(3 hours / person)} \cdot \text{(60 minutes / hour)}$$
$$\cdot \text{(10 ducks / minute)} \cdot \text{(2 laughs / 10 ducks)} \cdot \text{(4 shots / laugh)}$$
$$= 8.6 \times 10^{10} \text{ shots.}$$

That's 86 billion shots fired at the *Duck Hunt* dog. And yes, he deserved every bullet.

---

20 If you are a person who has played more than 20 hours of Duck Hunt, what is wrong with you? Seriously, do people do this?

The Plumber Jumped Over the Moon

Even before the *Madden* NFL video game series allowed you to create seven-foot-tall, 350-pound wide receivers, there were plenty of over-the-top video game studs to choose from: QB Eagles and Bo Jackson in *Tecmo Bowl*, Jeremy Roenick in *NHL 2003*, Mike Tyson in *Punch Out*.[21] The list goes on. Although these video game superathletes rightly get props for their hilariously unrealistic skills, there's one omnipresent video game character whose athletic prowess doesn't get nearly the respect it deserves: Mario. Originally nicknamed "Jumpman" in the 1981 video game Donkey Kong, the lovable Italian plumber has compiled quite an impressive athletic resume. At present, he's appeared in baseball, basketball, golf, and tennis video games, not to mention his side career as a racecar driver in *Mario Kart*. But his original moniker is perhaps the most suggestive indicator of Mario's most astounding athletic ability. **How high can Mario jump?**

It's difficult to discern exactly how high Mario can jump by just observing him during a game. At first glance you'd think he's only jumping over turtles and fanged mushrooms, which doesn't sound all that impressive. Whether or not these are mutants, though, is unclear, so perhaps we shouldn't assume they're the same size as the mushrooms and turtles in our common experience. A second thought would be to compare the height of Mario's jump to his own height, but even this is not so simple because whenever he eats a mushroom he doubles in size. It's unclear whether premushroom or postmushroom Mario is one who's supposed to be normal-sized, or even if neither is normal-sized. Be that as it may, Mario's own height is probably the most logical ruler to use. Even if it's not, we're talking about the vertical leap of a

---

21 For a nice list of other great gaming athletes, check out Scott Altman's "The 50 Greatest Video Game Athletes of All Time" on Bleacher Report, http://bleacherreport.com/articles/486314-the-50-greatest-video-game-athletes-of-all-time-with-video.

fictitious video game character, so I'm willing to take a certain amount of leeway in the believability of my assumptions.

Let me assume Mario is at full height after eating the mushroom so the results are at least somewhat more realistic. In *Super Mario Bros*, a large Mario can jump twice his own height. Assuming a 6-foot plumber, Mario's vertical leap would be 12 feet. Putting that into context, Mario could jump over Kobe Bryant sitting on Yao Ming's shoulders. He wouldn't even have to tuck his knees. Gerald Sensabaugh's NFL Combine record 46-inch vertical leap is one-third this height. In contrast to Mario's vertical leaping abilities, his long jump is somewhat pedestrian. At full speed Mario can jump about ten times his width. Assuming a 2-foot wide Mario, his long jump would be 25 ft. This is not a world record by today's standards, but it would have been back in 1922, so it's still fairly impressive. In any event, Mario's amazing leaping ability certainly dictates that he deserves a plaque in the video game athletic hall of fame.

Best. Goalie. Ever.

Never mistake a "hockey league for people who have never played hockey before" for a "hockey league for people who have never skated before." Concussions ensue. During my ill-advised attempt at athleticism on ice, I was struck by one question in particular. If we padded a morbidly obese guy and shoved him in the goal, could he block every shot? **How fat would a hockey goalie need to be to completely block the goal?**

Consider a cylindrical goalie.[22] Because an official NHL hockey goal is four feet tall and six feet wide, a six-foot tall goalie would be more than tall enough to cover the top of the goal. To span the width he must have a minimum radius of three feet. Using the formula for the volume of a cylinder, we can estimate his total volume:

$$\text{volume} = \pi \cdot (\text{radius})^2 \cdot (\text{height}) = 3.14159 \cdot (3.0 \text{ ft})^2 \cdot (6.0 \text{ ft})$$
$$= 4.8 \text{ m}^3 \ (\sim 170 \text{ ft}^3).$$

Some people can float in water while others sink,[23] so our bodies must have a density close to that of water, roughly 1000 kg/m^3. From this we can compute his total mass:

$$\text{mass} = (\text{density}) \cdot (\text{volume}) = (1000 \text{ kg/m}^3) \cdot (4.8 \text{ m}^3) = 4,800 \text{ kg}.$$

At over five tons, our heavyweight goalie would need to weigh about eight times more than the world's largest man.

---

22 Something about this statement makes me extraordinarily giddy.

23 This guy would probably float.

There's No "I" in STD

You don't have to be one of Tiger Woods's 19 mistresses to know that athletes love sex. In the professional sports world, athletes are much more likely to be a Tiger Woods than a Tim Tebow or an A. C. Green, but even by these heightened standards, one athlete stands out from the crowd. In his autobiography *A View from Above*, Wilt Chamberlain claimed to have slept with 20,000 women.[24] **How many STDs did Wilt Chamberlain contract?**

Chamberlain was about 55 when his autobiography came out. Assuming he started having sex at age 18, he would need to have sex with an average of 1.5 new women each day. There are a variety of STDs lurking out there. Some, like genital herpes, can infect as many as 1 in 6 people. Others, like HIV, are more deadly but less common, infecting about 1 in 700 people.[25] We can take the higher number because we know at least this many people are infected with an STD. This means that at least 1 out of 6, or roughly 3,300 of Wilt's "acquaintances," likely had an STD.

The rate at which STDs are transmitted depends on a variety of factors (disease type, overall health of the individuals involved, whether or not a condom was used, etc.). Wilt's serial philandering occurred largely before the discovery of AIDS, so he likely didn't practice safe sex. As a rough order of magnitude estimate, I'm going to assume a 10 percent transmission rate for unprotected intercourse between infected and uninfected individuals.[26] At this rate, Wilt would have contracted 330 STDs.

---

24 Little known fact: being tall and lanky wasn't the only reason Wilt was called "The Stilt."

25 Statistics obtained from Book of Odds at www.bookofodds.com.

26 For some diseases like HIV, the infection rate can be significantly lower.

Even if Wilt practiced safe sex 100 percent of the time, his sheer volume of partners still puts him at high risk. Condoms are advertised to be about 98 percent effective, meaning that 2 percent are not effective in preventing the spread of disease.[27] At this rate, even with condoms, Wilt would have contracted about 70 STDs.

---

27 It's unclear whether the 2 percent figure means that the condom wearer was infected 2 percent of the time or merely that he is as likely to be infected as he would be without wearing a condom. I am assuming the latter for simplicity.

Nobody Loves Me, Everybody Hates Me

"Bill Buckner tried to kill himself by jumping in front of a bus. Fortunately, it went through his legs."

It's not easy being the goat: Scott Norwood's kick sailing wide right, Chris Webber's untimely time out, Kyle William's muffed punt return. Even star athletes become the butt of tasteless jokes after one bad playoff game. While having a thick skin may ease the burden of having all that vitriolic hate spat at you, it won't necessarily ease the burden of having actual spit spat at you. **How many spitting spectators would it take to drown an athlete in drool?**

To solve this problem, we must first determine how much spit is required for drowning. In a sea of saliva, one could survive by swimming, but it's also possible to drown in your own bathtub. Let's take the volume of a bathtub as a lower bound. A typical tub might hold about 0.5 m³ of water. Let's assume each fan contributes 1 tablespoon of loogie. From this, we can find the total number of fans required to drown an athlete in spittle:

$$\text{\# of fans} = (0.5 \text{ m}^3) / (1 \text{ tablespoon per fan})$$
$$= 34{,}000 \text{ fans.}$$

If everyone in Fenway Park spat at once, you would have a bathtub's worth of saliva to drown yourself in. If you wanted a sea of saliva, you would need a lot more people. Even if the entire world population spat into Fenway Park at once, it would only reach halfway up the Green Monster.

Burning Rubber

As a kid, I was always confused when the treads on my high tops wore down. (As an adult, I now wonder the same thing about my car's tires). Where did they go? You might suspect that tiny chunks of material get rubbed off, but if so, where was the evidence? Shouldn't you find little bits of sneaker dust in the same way you find those worm-shaped pieces of pencil matter after erasing? After one step, you don't see any material left on the ground, and the sneaker bottoms look exactly the same. **How much rubber do you lose with each step?**

The treads on the bottom of sneakers can be about 4 mm deep. If you only use your shoes at the track, they might last about 6 months before the treads have worn down. A dedicated recreational athlete might run about 5 times per week. Between warm-ups and warm-downs, you might run about 2 miles per workout on average. Each running step transports you about 1 m. From these assumptions we can estimate thickness of rubber lost in each step:

$$\text{(4 mm / 6 months) / [(5 workouts / week)} \times \text{(2 miles / workout)} \times \text{(1 step / m)] = 10 nm/step.}$$

That's about one molecule's worth of material lost with each step. Seemingly miniscule amounts of wear and tear eventually add up. This phenomenon plays a large role in all sports. For example, the rate of wear is crucial in Formula One racing where the lighter tires are only built to last about 150 km, less the length of a race. More generally, any time an athlete moves, the material in his/her joint degrades by a small amount. While biological tissues have the ability to repair themselves, this ability is finite. Eventually, the rate of wear surpasses the rate of repair. Biological degradation of this type is accurately described by the colloquialism, "Hey, it sucks getting old."

Don't Stash the 'Stache

Facial hair has a long history in sports. From Rollie Finger's handlebar mustache to Hulk Hogan's Fu Manchu to whatever the hell you call that thing on Kimbo Slice's face, an athlete's facial hair can become almost as iconic as the athlete himself. Recently, facial hair has taken an even more prominent role with playoff beards becoming all the rage, despite the fact that the added itchiness would seem to create a competitive disadvantage. Originally only an NHL tradition, the ritual has spread to other major sports with Ben Roethlisberger and LeBron James being two of the more notable practitioners. Given the seemingly limitless number of styles to choose from, a playoff-bound fashion-conscious athlete is faced with a dilemma. "Should I go with the Brian Wilson hirsute Spartan or settle for the Jayson Werth mountain man?" Physicists aren't exactly known for their fashion sense, but I'm comfortable offering this little tasty advice nugget: If all you can muster is three scraggly hairs poking out of your chin, don't call it a playoff beard. I'm looking at you, Sid Crosby. **Just how many potential beard combinations are there?**

To start, it's helpful to calculate the number of hairs in one's beard. A typical adult male might have facial hair covering an area roughly 30 cm by 10 cm, or about 300 cm^2. If each hair were separated by 1.0 mm, then he would have about 30,000 hairs. With this number in hand, we must now answer a difficult question. Just what differentiates one beard from another? (Later in the book, we'll discuss something called "order parameters." For the present problem, it would help to have a beard order parameter.)

For example, there comes a certain point when one exits Brady-Anderson-sideburn land and enters the realm of the Jonathan Toews' mutton chop. This is not a sharp line; by simply plucking one hair you don't unambiguously transition from mutton chop to sideburn. There's some fuzziness here (pun intended), but it's our job to make a rough estimate.

Let's assume one can visually distinguish when 1 cm^2 of hair has been removed. With this criterion, we can grid up one's face into 300 square-centimeter patches that can either be filled with hair or not filled with hair. Since there are only two options (i.e. hair or no hair), we can easily estimate the number of possible beards (I should note that I'm neglecting hair length, which would lead to many more beard types.):

$$\text{\# of beards} = 2^{300} \approx 10^{90}.$$

That is a huge number of beards. It's vastly bigger than the number of atoms that could fit in our solar system. Why so many beards? Well, it turns out no one in their right mind would ever attempt to try the vast majority of these beards. Our beard list includes everything from checkerboard patterns to beards where you have a goatee on the left side of your face and a neck beard on the right. Humans tend to like symmetry, but if we're enumerating the number of possible beards rather than the number of halfway decent looking beards, then we have to include the ones even Dennis Rodman would think were too weird. However, if you're looking for a challenge, see if you can estimate the number of beards one might actually try. (Warning: This a much more difficult problem!)

SCALING THE RUNGS OF LITTLE LEAGUE

I live in a world where h-bar is one. The speed of light is one. Pi is one.

—**Physicist Edward Farhi during an undergraduate quantum mechanics lecture**

On the whole, physicists tend to be more interested in functions than absolute numbers. Viewed in this light, numerical constants are like the cat food–shaped oat pieces in an otherwise marshmallow-filled bowl of Lucky Charms. In short, they just get in the way. This idea can be a bit jarring to both laypeople and introductory physics students, who are used to calculating actual numbers. What can be particularly unnerving is that many times it's not just constants like h-bar and the speed of light that are set equal to one; I distinctly recall several instances when 2 was equal to 1. Although it may seem like we're going out of our way to fit the stereotypical image of a wild-haired physicist who has lost touch with reality, I assure you that we only do it to see functional dependencies more clearly.

We're all familiar with the basic idea behind functions. The very word permeates our lexicon. If I were to ask you what winning was a function of, you might say hard work, dedication, skill, and so forth. This is true mathematically because the probability of winning a sports contest increases as each of these variables increase. It can equally be said that winning is a function of how much alcohol you consume, how much cocaine you snort, and how many prostitutes you invite to your hotel room the night before a game.[28] Unless you're Lawrence Taylor, it's very likely that increasing any one of these variables will severely decrease your probability of winning. It doesn't matter if it's something as noticeable as the number of practice hours or as overlooked as the athlete's shoe size. If it affects the outcome of the game, then

---

28 In terms of notation, you write the symbol for the function first followed by the symbols of all the variables in parentheses. For example, the probability of the winning function described in the text could be written as $P(a, c, p)$, where a is the amount of alcohol consumed, c is the amount of cocaine snorted, and p is the number of prostitutes hired.

winning is a function of it. When a physicist says he's interested in functional dependence, he means he wants to know how some variable (e.g., probability of winning) changes as you change other variables (e.g., hard work, dedication, cocaine snorted, etc.).

Imagine an owner increases the size of a stadium by 10 percent to help reduce his pitchers' earned run average. This is equivalent to moving the foul poles from 300 to 330 feet. How much more area do the fielders have to patrol? The natural response is to say 10 percent more, but this is incorrect. It's actually 21 percent more. Consider the following:

We've increased our length L by 10 percent to a new length L'. This 10 percent increase means that L' will be 1.1 times the size of the original length L. This can be written symbolically as

$$L \to L' = 1.1\,L$$

Area is calculated by squaring a length, so we could say that the area A of the playing field scales is L^2. We write this statement symbolically as

$$A \sim L^2$$

where the symbol "$\sim$" means "scales as." Using these symbols, we can solve for how much the area A' increases:

$$A \to A' \sim L'^2 = \left(1.1\,L\right)^2 = 1.21\,L^2 \sim 1.21\,A$$

As you can see, the area of the field increases by 21 percent. Hitters may have to hit the ball 10 percent further for homeruns, but now fielders have to cover 21 percent more ground. Although you reduce the number of home runs hit, that doesn't necessarily compensate for the increase in singles, doubles, and triples.

This is an example of a scaling argument. It's a quick and dirty way to do math because you don't have to solve equations. You can just think about how certain quantities scale with (i.e., how certain quantities are functionally dependent on) other quantities.

400m Marathons

If you've ever watched a track meet, you may have noticed that for events longer than 100 meters, runners in the outer lane get a little bit of a head start. The reason is fairly simple: the outer lanes are longer. The curved section of the outer lanes is basically a semicircle. Because the circumference grows as the radius increases, the outer lanes must be longer by simple geometry. **How many lanes would you need for the outermost lane to be the length of a marathon?**

The distance around a standard track is 400 meters, whereas a marathon is 42 kilometers (~26.2 miles). The radius of curvature for the curved section of the innermost lane is about 30 meters. Although the length of the curves increases the farther out you get, the straightaways are 100 meters no matter which lane you're in. Because the total length of each straightaway is much smaller than a marathon, the vast majority of our hypothetical track is curved, so to a very good approximation, you're running in a circle. For this reason, I'm going to neglect the straightaways and assume a circular track. The circumference C of the outer edge of the circular track is 42 kilometers, and the inside lane, from which we've removed the straightaways, is now 200 meters. This means the outermost lane needs a circumference 210 times larger than the innermost lane. From the formula for the circumference of a circle, we can see that the radius R scales linearly with the circumference C:

$$R \sim C$$

Denoting the radius and circumference of the outermost lane by R' and C' respectively, we can see that

$$R' \sim C' = 210 \, C \sim 210 \, R$$

The radius of the outer lane will be 210 times larger than the inner lane, or roughly 6.3 kilometers (~3.9 miles). Because the width of an athletic track is about 48 inches, you would need about 5,200 lanes if you wanted the outermost lane to be a marathon in length.

The Pressure's On

Running up and down the NBA hardwood for 80 games a year, ten years straight can be hell on your knees. This is particularly evident in NBA big men. Consider the difference between Shaquille O'Neal and Mugsy Bogues. Shaq tips the scales at 325 pounds, whereas Mugsy weighs in at only 136 pounds, so it's tempting to conclude that any excess knee damage in NBA centers is purely due to the extra mass. Strictly speaking, however, this is not correct. Elephants can support 100 times as much weight as humans do, despite the fact that their bodies are composed of fundamentally the same types of fat, proteins, and other biomolecules as human bodies, so there has to be more to it than just the net weight. The reason elephants can support a larger body mass is partly because their legs are bigger in comparison to their bodies than a human's legs are. In contrast, a tiny animal like an ant has small legs compared to its body size. The quantity of interest, then, is not simply the weight on the knees but also how the weight is distributed over the knees. Wide legs can support more weight because the weight will be spread out. Since pressure is defined as force (or weight) per unit area, this is a more useful quantity to consider.[29] **How much larger is the pressure in Shaq's knees?**

At 7 feet, 1 inch, Shaq is roughly 1.35 times taller than the 5-foot-3 Bogues. Assuming they're proportioned similarly, the radius of Shaq's knees would likewise be 1.35 times larger. The cross-sectional area A of the players' knees will scale with the knee radius R as

$$A \sim R^2$$

From this, we find that the cross-sectional area of Shaq's knees is

$$A_{Shaq} \sim R_{Shaq}^2 = \left(1.35\, R_{Bogues}\right)^2 = 1.82\, R_{Bogues}^2 \sim 1.82\, A_{Bogues}$$

---

29 We'll discuss pressure in greater detail later in the book. For now, this will suffice as a definition, but at the risk of being pedantic, I should note that physicists generally distinguish between weight, which is a force, and mass, which measures the amount of matter in an object.

Because Shaq weighs about 2.4 times more than Bogues, the pressure in his knees is

$$P_{Shaq} = \frac{W_{Shaq}}{A_{Shaq}} = \frac{2.39\, W_{Bogues}}{1.82\, A_{Bogues}} = 1.31\, P_{Bogues}$$

That's about 31 percent more pressure on Shaq's knees. You might say, "Well, yes, but Shaq's a fatty." What happens if you scale Mugsy Bogues up to Shaq's size?

To see what happens as we increase Bogues's height to that of Shaq's, you have to increase all his lengths by a factor of 1.35 just to keep everything proportional:

$$L_{Bogues} \rightarrow L_{Bogues}{'} = L_{Shaq} = 1.35\, L_{Bogues}$$

Because weight W scales with volume V, which scales as length cubed, we can write Mugsy's new weight, $W_{Bogues}{'}$, as

$$W_{Bogues} \rightarrow W_{Bogues}{'} \sim \left(L_{Bogues}{'}\right)^3 = \left(1.35\, L_{Bogues}\right)^3 = 2.46\, L_{Bogues}^{3} = 2.46\, W_{Bogues}$$

This means a 7-foot-1 Mugsy would weigh 334 pounds. Once you correct for the height difference, Mugsy actually weighs more than Shaq does, so the extra pressure in the Big Diesel's knees can't be attributed to having a little extra baggage. Unless your 7-footer has unusually wide knees, he's necessarily going to have a lot of pressure built up inside them. When you're that tall, every day is like running up and down the hardwood floor.

Honey, I Shrunk the Pisarenko[30]

It's commonly said that ants can hold 50 times their own weight, but is this really as impressive as it sounds? After all, they certainly can't pick *us* up.[31] If we want a fair comparison, we need to derive how much weight each species could lift if they were equal in size. **How much weight could weightlifter Anatoly Pisarenko lift if he was ant-sized?**

As mentioned in the previous problem, ants have tiny legs compared to their body size, whereas humans have medium-sized legs and elephants have huge legs. The reason for this is that bigger species must support more weight. Just like basketball players, if you double the height of an animal but keep its limbs in proportion, then it will be two times as tall, but it will weigh $2^3 = 8$ times as much because its weight scales as the length cubed. Because the force holding up the weight is proportional to the area, it will grow as the square of the length and be $2^2 = 4$ times as large. If a human were to shrink down to the size of an ant, the amount of weight he could lift would shrink, but his own weight would shrink even faster.

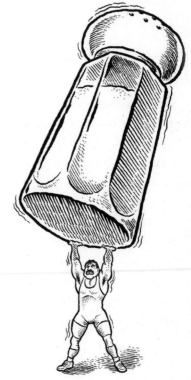

---

30 I admit that I had some difficulty choosing between Pisarenko and Naim Süleymanoğlu, but I couldn't justify shrinking an already pocket-sized Hercules.

31 In my book *How Many Licks?: Or, How to Estimate Damn Near Anything*, I estimate that it would take 65,000 ants to hold up one human.

Let's say Pisarenko weighs 100 kilograms (~220 pounds) and is about 1.8 meters (~6.0 feet) tall. If we shrink him down to the height of an ant, he'll be about 3 millimeters tall, or roughly 600 times shorter. Using the fact that volume scales as the cube of the length, we can calculate his new mass:

$$m \rightarrow m' \sim L'^3 = \left(\frac{1}{600} L\right)^3 = 4.6 \times 10^{-9} L^3 \sim 4.6 \times 10^{-9} m$$

That's about 0.46 milligrams. Pisarenko clean and jerked 258 kilograms (~580 pounds). Using the fact that the weight that can be supported scales roughly as the square of the length, we can estimate how much weight his new tiny self could pick up:

$$W \rightarrow W' \sim L'^2 = \left(\frac{1}{600} L\right)^2 = 2.8 \times 10^{-6} L^2 \sim 2.8 \times 10^{-6} W$$

which translates to about 716 milligrams. Comparing these two numbers, we see that an ant-sized Pisarenko could lift an astonishing 1,500 times his own weight. If you want to play around with these scaling relations, try estimating how much weight a human-sized ant could lift. Is there any chance that giant ants will take over the world?[32]

---

[32] You should be able to show that a human-sized ant would not be strong enough to lift itself.

Scaling in the Universe

Go to YouTube and search for the phrase "moon golf." You'll find a video of astronaut Alan Shepard on the Moon using a six-iron to hit a golf ball. Even though the bulky spacesuit forced Shepard to swing with one hand, the ball still traveled, in Shepard's words, "miles and miles and miles."

Celestial bodies differ in their gravitational strength. At the surface of the Earth, objects accelerate downward at a rate of approximately $g = 9.8$ m/s². **How far could you hit a golf ball on each of the major bodies in our solar system?**

As you'll see when you come to the section on projectile motion, the range (i.e., distance traveled) of a projectile scales as

$$R \sim \frac{1}{g}$$

where R is the range and g is the gravitational acceleration on whatever celestial body you happen to inhabit. Consider hitting a golf ball on the Moon. The Moon's gravitational acceleration is about 16 percent of Earth's gravitational acceleration, so we can replace Earth's value of g with $g' = 0.16\,g$. The range R' of a golf ball on the Moon can be determined using scaling:

$$R' \sim \frac{1}{g'} = \frac{1}{0.16 \cdot g} = 6.3\,\frac{1}{g} = 6.3\,R$$

A drive on the Moon will travel 6.3 times farther than on Earth. For example, a 300-yard drive will now travel over a mile. Using the same scaling relationship along with the surface gravities for the Sun and other planets, we can compute how far a 300-yard drive will travel in each case. The results are shown in the table on the next page.

	SURFACE GRAVITY (IN EARTH GRAVITIES, g)	RANGE (IN YARDS)
SUN	28	10
MERCURY	0.38	789
VENUS	0.903	332
EARTH	0.16	1,875
MOON	1	300
MARS	0.38	789
JUPITER	2.53	118
SATURN	1.07	280
URANUS	0.92	326
NEPTUNE	1.12	267
PLUTO	0.06	5,000

As a bonus, here's how far Tom Dempsey's record-setting, 63-yard field goal would have traveled on each of the heavenly bodies:

	SURFACE GRAVITY (IN EARTH GRAVITIES, g)	RANGE (IN YARDS)
SUN	28	2
MERCURY	0.38	165
VENUS	0.903	69
EARTH	0.16	393
MOON	1	63
MARS	0.38	165
JUPITER	2.53	24
SATURN	1.07	58
URANUS	0.92	68
NEPTUNE	1.12	56
PLUTO	0.06	1,050

Apparently, Dempsey's kick would only travel two yards on the Sun, but it would be good from almost a kilometer away on Pluto.

Buzzer Beaters: Or, When Scaling Goes Bad

One of the most thrilling plays in sports is basketball's full-court buzzer beater. It's also one of the most rare events in sports. **What's the probability of hitting a buzzer beater?**

Before jumping straight to buzzer beaters, it's helpful to first consider the simple case of free throws. To make a free throw, the shooter, standing 15 feet away from the basket, must toss the ball with a speed and angle such that the ball passes through a hoop, which has some area A. Mechanically speaking, it doesn't matter how he maneuvers his limbs so long as the end result is that the ball leaves with a speed and angle that will make it go through the rim. When a player misses his second free throw after making his first, you can be sure that the ball left with a different speed or angle. If you could reproduce the same motion every time, you'd be a 100 percent free throw shooter, but any small fluctuations in your body's motion can send the ball off course enough to miss. Even a mistimed heartbeat can be the difference between making the NBA finals and going home early.

A decent free throw shooter might hit only 80 percent of his shots.[33] Michael Jordan's free throw percentage was 83.53 percent, but the all-time record belongs to Mark Price at 90.39 percent. This means that 90.39 percent of Mark Price's shots landed in a small area A located ten feet above the hardwood floor.

What would happen if Mark Price had to shoot from a farther distance? He'd certainly have to shoot the ball with a higher speed to increase the distance, but he wouldn't necessarily have to change the angle. Let's, for the moment, make the naive assumption that the angle of the shot remains the same for a longer shot, but the speed is adjusted to correct the distance. If

---

33 You might say that free throws aren't defended, so those are easier to make. True, but full-court shots aren't typically well guarded either.

that were the case, what would happen to Price's shot percentage as we increased the distance between him and the rim? Well, the angle stays the same, so you might think the shot percentage does as well, but when shots are fired from a longer distance, shooting percentage generally decreases. Imagine a movie projected on a wall. If you move the projector back farther, the image gets larger. Similarly, if you move the shooter back farther, the "image" of the where the shots land gets larger. Shots that previously landed within the area A are now landing in a substantially larger area A'. You can see this by noting that the farther away you get from the rim, the smaller it looks. Using a little geometry, it's not hard to see that the area A in which shots land scales with distance as

$$A \sim L^2$$

For a full-court buzzer beater, the shot must be taken from about 90 feet, which is 6.3 times longer than the distance from the free throw line. This means we're looking at a change in distance,

$$L \to L' = 6.3L$$

All of the shots that would have gone in before are now spread out over an area A' that scales as

$$A' \sim L'^2 = \left(6.3\,L\right)^2 = 40\,L^2 = {\sim}40A$$

Only a fraction of shots that would have gone in before will go in now. Assuming the shots were uniformly distributed over the area of the rim originally, the fraction of them that go in now will be

$$\left(\text{original shot percentage}\right) \cdot \left(\frac{A'}{A}\right) = \left(90.39\%\right) \cdot \left(\frac{1}{40}\right) = 2.3\%$$

That's a small percentage to be sure, but it's larger than you'd expect in reality. There are 1,230 regular season NBA games played each year. If players attempted 1 full-court buzzer beater per game, you'd see almost 30 per year. Moreover, you can easily see this can't be correct by noting that any shot, even one from 37 miles away, has a finite probability of landing in the net. Even Michael Jordan's not that good.

Where did we go wrong? First off, we assumed the accuracy of the shot doesn't depend on how hard it's thrown, which is not at all realistic. Common sense tells us that being accurate is a lot easier if you're not trying to gun it. Furthermore, in order to apply the extra force needed to shoot the ball from a farther distance, you're necessarily altering your shooting form, which will affect your accuracy. The moral of the story: scaling arguments are great, but they're not always applicable when the situations are too complex. Don't be afraid to use them in simple cases, but always be wary whenever you're applying scaling to complex systems like the motion of a human body.

WELCOME TO VARSITY

If I asked you to complete the lyric, "Take me out to the _____," there's little doubt you'd say anything but "ball game." It's somewhat surprising, then, that when confronted with an equation like $3x^2 - 2 = 4$, people are often overcome with dread. Algebra is nothing more than a mathematical *Mad Lib* in which you fill in the blanks with numbers rather than words.[34] To be sure, algebra is more difficult than a typical *Mad Lib* because the numbers you plug in must be consistent with what's going on in the equation. Still, the basic concept is the same, and if you're one of the many people who left school feeling uneasy whenever numbers got mixed up with letters, then hopefully this analogy will provide some perspective to lessen the burden. In any event, now is the perfect time to overcome your fear because we'll be using algebra extensively for the rest of the book.

Algebra makes use of the fact that expressions on either side of an equal sign are equivalent. Take Henrik Sedin and his genetic clone Daniel. If we assume the hockey twins are identical in every way, then we can stick an equal sign between them. This equality will hold as long as we do the exact same thing to Daniel that we do to Henrik. If we surgically remove Daniel's pinky fingers, the two will still be identical provided we do the same to Henrik. Algebra follows the same rules, but we're usually interested in solving for a particular variable. To do this, you want to isolate this variable on one side of the equal sign. For example, you could solve for Henrik's left big toe by removing all his other body parts until only the toe remains and then doing the same to Daniel. I've clearly overused my organ harvesting analogy at this point, but you get the idea. Algebraically, you isolate a variable by undoing all the operations on the variable. For example, if you want to isolate the *x* in

$$x + y = z,$$

---

[34] For unfamiliar readers, a typical Mad Lib might go as follows: think of the name of a baseball player, a body part, and a plural noun. Now enter them into the following sentence: "Jose Canseco took \<name of baseball player> into the bathroom and injected his \<body part> with \<noun>."

you have to undo the "+ y" by subtracting y from both sides. You can do this because subtraction is the opposite of addition. Likewise, you can undo multiplication by dividing. How do you undo exponentiation? It's a trick question because there are two answers. If you want to solve

$$x^y = z,$$

for x, you take the y-th root of both sides. However, if you instead want to solve for y, you have to take the logarithm of both sides and then divide by log(x). The reason there are two "opposites" for exponentiation is that it is not commutative (i.e., x^y is not the same as y^x). By undoing operations until you're left with the variable you want, you should be able to solve most of the algebraic problems in this book.

Unfortunately, doing good physics often involves more than algebra. In many cases, calculus is absolutely essential. As I mentioned before, I'll try to keep the calculus to a minimum, but in the rare instances when it's necessary, don't be frightened. Calculus is basically just adding and dividing a bunch of really small numbers. You can use it to compute how a function increases, how much space it occupies, and where it's largest. If this scares you, remember that we use calculus in the same way we use algebra except instead of solving for a variable x, you solve for a function f(x). In other words, calculus is just another mathematical *Mad Lib*.

Because many equations will be introduced in this section, I've divided this chapter into eight sections based on the type of physics involved. At the beginning of each section, I give a brief description outlining the physics you need to know and describing the equations.

PHYSICS I:
The Speed Department

In sports we're mainly concerned with bodies in motion. As such, it's helpful to construct a mathematical framework from which we can describe an object's motion via its position $x(t)$ as a function of time t. For example, $x(t)$ might represent how high off the ground a baseball is at any moment in time. Ideally, the framework for characterizing this function should include descriptions—or at least approximate descriptions—of any arbitrary motion that is physically realistic. Physicists have developed such a framework, and it's very good for describing the motion of baseballs, footballs, and hockey pucks.

To begin, we need to know where the object is initially. This is easy enough because we can always measure its position. Now, if we only measured where objects were, life would be pretty boring. Objects can move around from one place to another, and when they move, they do so at a certain rate. This rate is called velocity, and it can be measured just like position.

Now, if all we did was measure where objects were and how fast they were moving, life would be pretty boring. Objects can change velocity while going from one place to another, and when they change velocity, they do so at a certain rate. This rate is called acceleration, and it can be measured just like position and velocity.

Now, if all we did was measure where objects were and how fast they were moving and how fast their velocity was changing, life would be pretty boring. Objects can change accelerations while going from one place to another, and when they jerk around like this, they do so at a certain rate. This rate is called jerk, and it can be measured just like position, velocity, and acceleration.

Now, if all we did was measure where objects were and how fast they were moving and how fast their velocities were changing and how much their acceleration evolved, life would be pretty boring . . .

You're probably starting to detect a pattern. I'll spare you the details, but we can continue this indefinitely. The next few terms in the series after the jerk have jokingly been named the snap, crackle, and pop, but after that, no one really pays attention. In fact, most physicists stop caring after acceleration. We do this for a very fundamental reason: it makes equations easy to solve. It turns out acceleration also has an important physical property, but I'll talk about that when we get to forces. If we ignore everything after the acceleration, we get an equation that looks like this:

$$x(t) = x_0 + v_0 t + \frac{1}{2} a_0 t^2$$

where x_0, v_0, and a_0, are the initial values for position, velocity, and acceleration, respectively.[35] The subscript "0" (pronounced "not") next to each of the symbols is there to specify that the value is taken initially when $t = 0$.

35 You may wonder where the ½ in front of the acceleration comes from. After time t, your velocity will increase by an amount at, but on average it will have increased by ½at. If you combine this with your original velocity v_0, you arrive at exactly the equation shown in the text.

Also, as before, you may detect a pattern emerging. Each term in the series has the variable t raised to some power divided by a number. A moment of reflection will reveal that the Nth term in the series has t to the Nth power divided by all the integers from 1 to N multiplied together. Including more terms results in an equation,

where j_0, s_0, c_0, and p_0, are the initial values of jerk, snap, crackle, and pop, respectively.
There's something I should warn you about. If you took high school physics, you may have been duped into thinking there's some fundamental physics going on here. There's not, save for the fact that the equation implies that objects move continuously as opposed to jumping discontinuously from one point in space to another. Other than that, this is just a mathematical approximation that describes motion. It is not a physical law in the sense that it doesn't tell us anything about how the universe works. If you doubt this, let me point out that you can approximate other functions ($v(t)$, $a(t)$, etc.) using exactly the same form. Approximating a function using a power series like this is called a Taylor series after mathematician Brook Taylor.

This equation may look familiar if you took high school physics. It's the equation of motion for an object moving with constant acceleration. The assumption of constant acceleration is reasonably good for many sports-related phenomena, including baseballs pulled down by gravity and sprinters accelerating out of the blocks.

With constant acceleration, the velocity at any moment in time can be found using the equation:

$$v(t) = v_0 + a_0 t$$

Congratulations! Without realizing it, you've just had a calculus lesson. Although I've glossed over the details, we're actually talking about derivatives. A derivative is basically fancy math-speak for looking at how a function changes as you change one of its variables. More specificially, a derivative is the rate of change of a function with respect to some variable. In this case, the derivative in question is velocity, which is the rate of change of position with respect to time. If we look at how our position function changes over some small amount of time, we can approximate these derivatives as[36]

$$v = \frac{\Delta x}{\Delta t}$$

where x is how much the position changes during the time interval t.

In truth, we actually considered two derivatives: velocity and acceleration.

---

36 The funny looking triangle symbol "Δ" is a Greek letter pronounced "delta." It is a general symbol that physicists and mathematicians use to denote a change in a physical quantity. For example,

$$\Delta x = x(t) - x_0$$

is the change in position that happens during a time, and

$$\Delta t = t - t_0 = t$$

where I've defined the initial time t_0 to be zero.

The acceleration is the rate of change of velocity and can be written as

$$a = \frac{\Delta v}{\Delta t}$$

where v is the change in velocity that occurs during a time interval t.

With these equations in hand, you now have enough information to solve a variety of physics problems. Your challenge now is to figure out how to solve the questions that follow by rearranging the equations above and estimating the values you should plug into them.

Crack of the Bat

As a physicist, I often get asked how we know certain things: the speed of sound or light, the mass of the Earth, the size of an atom, and so forth. Here's a neat trick you can do to calculate the speed of sound the next time you're at the ballpark.

If you've ever sat way back in the centerfield bleachers, you may have noticed a peculiar effect during the game. There seems to be a slight time delay between when you see the ball hitting the bat and when you hear the ball hitting the bat. The reason is that it takes more time for sound to reach you than it does for light.[37] **Using your ballpark experience, how fast do you think the speed of sound is?**

If you're like me, you always get the cheapo seats, which can be about 500 feet away. From this vantage point, you generally hear the crack of the bat anywhere from 0.2 to 1 second later, so I'll assume the delay is about 0.5 second.

Upper Bound: 1.0 s
Assumed Average: 0.5 s
Lower Bound: 0.1 s

From this we can estimate the speed of sound:

$$v = \frac{\Delta x}{\Delta t} = \frac{(500 \text{ ft})}{(0.5 \text{ s})} = 300 \text{ m/s}$$

---

37 It's the same reason you see lightning before you hear thunder. The next time you're in a thunderstorm, count the number of seconds between lightning and thunder and multiply by 0.2 to estimate how many miles away the strike was.

According to our estimate, the speed of sound is about 300 m/s. Our bounds indicate that the exact number should lie between 150 and 750 m/s. In the atmosphere the speed of sound is about 340 m/s, so our estimate is very close. Here, we've figured out a fundamental fact of nature using nothing more than our wits and a cheap ballpark ticket.

Ready-eady-eady, Set-set-set, Go-go-go,

Now that we know the speed of sound, we can solve a problem that's always bothered me. At a track meet the official fires the starting pistol signaling the runners to begin. The sound of the gun will reach each runner at different times depending on where the official is standing. **How much of a disadvantage do sprinters farthest away from the referee have?**

A quick web search shows that a standard athletic track has eight lanes, each of which is 48 inches wide. From this, we can conclude there are about 8.5 meters between the first and last sprinter because the two are separated by seven lanes. Using this and the speed of sound, we can calculate the time it takes for sound to travel from the first runner to the eighth:[38]

$$\Delta t = \frac{\Delta x}{v} = \frac{(8.5 \text{ m})}{(340 \text{ m/s})} = 0.025 \text{ s}$$

Usain Bolt's record-setting 100-meter dash time was measured at 9.58 seconds, so this time delay is significant and could be the difference between winning and losing a race.

---

38 I've assumed the referee is standing directly to the side of the runners. If he stands centered behind or in front of them, the sound will reach runners at opposite ends simultaneously but will reach the runners in the middle first.

Lightning Speed

It's been said Jackie Robinson was so fast he could flip the light switch and be in bed before it got dark.[39] Unless he had a peculiar light switch, this is almost certainly an exaggeration.[40] **How fast would Robinson have to run in order for this statement to be taken literally?**

To see how fast Robinson would need to be, we first need to know how fast electricity travels through the wires. Unfortunately, the "speed" of electricity can mean different things. We could be talking about the average speed at which individual electrons travel, but it turns out this is only a few millimeters per second.[41] If this speed determined how quickly lights turned on, you'd have to wait hours after flipping the switch before you could see anything. Because electrons throughout the wire are all moving together, light bulbs turn on almost instantaneously. For this reason, we're interested not in the speed of the electrons themselves but rather the speed of the signal that tells the electrons farther down the wire to move forward. Electrical signals travel at the speed of light. In a copper wire, light travels around 90 to 99 percent of the speed of light in vacuum. In vacuum, the speed of light is given by

39 The same statement has been made about Cool Papa Bell, Muhammed Ali, and othe fast athletes

40 In principle, a light bulb can be kept on for a substantial amount of time if it has a large inductance. This is what keeps those "shake-weight" flashlights glowing after you stop shaking them.

41 This speed is called "drift velocity." An individual electron moves slowly on average because it's constantly bouncing off nuclei and spends almost as much time moving backward as it does forward.

$$c = 2.9979 \times 10^8 \text{ m/s,}$$

so Jackie Robinson would need to run at least $0.9c = 2.7 \times 10^8$ m/s to beat the darkness.

How fast is this? In baseball there are 360 feet around the bases, and a typical ground out might happen in 4.0 seconds. If Robinson could run at $0.9c$, then he could run around the bases

$$\text{\# of times} = \frac{\Delta x_{travelled}}{\Delta x_{bases}} = \frac{v\Delta t}{\Delta x_{bases}} = \frac{\left(2.7 \times 10^8 \text{ m/s}\right) \cdot \left(4.0 \text{ s}\right)}{\left(360 \text{ ft}\right)} = 9.8 \times 10^6 \text{ times}$$

That's about 10 million times around the bases in the time it takes to ground out.

Reaction Time and Neuron Speed

Making good contact with a round ball and round bat even if you know what's coming is hard to do. That seems to be the one major thing that all young players have difficulty with. Why? It's the hardest thing to do in sports. So that's the reason baseball is a hard game to play.

—**Ted Williams**

Ted Williams was right. Baseball players have as little as 0.4 seconds to make contact with a sphere moving at speeds up to (and in some cases over) 100 mph. Whether you're a goalie making a save or a sprinter coming out of the blocks, fast reaction time is crucial for success in sports. But just how fast can a person be? **What is the fastest reaction time possible?**

Reaction time is a complicated beast. First off, visual information in the form of electromagnetic waves travels from the outside world through the lenses in your eyes, which constantly adjust to keep an image of the light focused on a small group of cells in the back of your eyeball. These cells contain molecules that absorb the light. When the light strikes the molecules, they twist about and rearrange themselves, and a byproduct of this rearrangement is that an electrical signal gets sent down some neurons into your brain. At this point, what was previously a bunch of electromagnetic fields wiggling in space has been converted to a bunch of electrons and ions wiggling in your brain. Just like a computer, your brain recognizes these moving charges as input. Of course, by "recognize" I really mean that the moving charges in your brain cause other charges in a different part of your brain to start zigzagging in a big complicated pattern, but the only thing you're consciously aware of

is "CURVEBALL!!!" With this information now interpreted, the jiggling electrons and ions in your brain start causing other electrons and ions in your
spinal cord to shimmy about, which in turn causes electrons and ions in your
nerve cells to fidget around, which will no doubt cause electrons and ions in
your muscles to reverberate in such a way that individual muscle fibers will be
pulled closer together, resulting in you swinging and missing at strike two. The
most remarkable part of all this is that it's all controlled by the laws of physics,
and it happens in under 0.4 seconds.

The whole process is fabulously complicated, and if we're going to have
any hope of calculating a reaction time, we need to simplify it drastically. That
said, there are two main causes that slow your reaction time: the time it takes
for your brain to process signals and the speed at which signals travel from
your brain to your muscles. Quick reaction time is as much a function of brain
processing speed as it is of muscle quickness. If your brain were running
Windows 95, your dream of being an elite NHL goalie would be DOA.
Fortunately, billions of years of evolution have made it so we can process
visual information very quickly and respond to stimuli almost as rapidly.
Unfortunately, the computer analogy only takes us so far, and as best I can
tell, there's not really a good way to calculate a brain's processing speed
a priori. That said, we still have to worry about the time it takes a nerve
impulse to get from point A to point B, so we can at least put some bounds
on reaction time. As an upper bound, we know humans can react in, at most,
0.4 seconds, because baseball players do this routinely. For a lower bound it
helps to look up the speed of neuron. According to a Wikipedia article on
"Action potential," neuron speed values range from 1 to 100 m/s. Taking the
larger of the two, we can compute a lower bound for reaction time by
assuming neurons have a length of about 1 meter:

$$\Delta t = \frac{\Delta x}{v} = \frac{(1.0 \text{ m})}{(100 \text{ m/s})} = 0.01 \text{ s}$$

The absolute fastest reaction time would be about 0.01 second. This would seem to be a fundamental limit. It doesn't matter how much you train, if your neurons can't send information to your limbs any faster, then you're pretty much stuck. Using fairly complicated equipment to measure the electromagnetic response of the brain, scientists have measured stimulus reaction times on the order of 0.08 seconds, which is within an order of magnitude of our lower bound.[42]

---

42 See, for example, David F. Griffing's The Dynamics of Sports: Why That's the Way the Ball Bounces, 3rd ed. (Oxford, OH: The Dalog Company, 1987).

I've Got the Need for Speed

During the 2008 Beijing Olympics, Usain "The Aptly Named" Bolt broke the 100-meter dash record with a time of 9.72 seconds. If the time wasn't impressive enough, Bolt could have finished faster had he not slowed up to celebrate before reaching the finish line. Bolt later shattered his record in Berlin with a time of 9.58 seconds. Although Bolt is clearly the fastest man on the planet, his speed hardly compares to that of man-made machines. **How much of a lead would Bolt need to beat racecar driver Dale Earnhardt Jr. in a 100-meter dash?**

For simplicity, I'm going to ignore the fact that both the runner and car have to accelerate. We can find Bolt's average speed in his record-setting 100-meter dash by dividing the distance by time:

$$v = \frac{\Delta x}{\Delta t} = \frac{(100 \text{ m})}{(9.58 \text{ s})} = 10.4 \text{ m/s}$$

That's roughly 23 mph. In contrast, NASCAR cars routinely reach speeds over 200 mph, meaning Earnhardt would finish the race in

$$\Delta t = \frac{\Delta x}{v} = \frac{(100 \text{ m})}{(200 \text{ mph})} = 1.12 \text{ s}$$

This means Bolt only has 1.12 seconds to complete the race. From his speed, we can compute how far he could travel in that time:

$$\Delta x = v \Delta t = (177 \text{ mph}) \cdot (1.12 \text{ s}) = 89 \text{ m}$$

Bolt would only travel about 10 meters, which means he would need to be spotted about a 90-meter lead in order to beat Earnhardt in a 100-meter race.

Bolt Physics

In 2010, Chris Johnson of the Tennessee Titans challenged Usain Bolt to a race. The NFL's fastest man, Johnson ran a 4.24-second 40-time at the NFL Combine. This got me wondering whether there was anything us physicists could predict about who would win in a 100-meter race. **Who would win in a race, Chris Johnson or Usain Bolt?**

Naively, one could compute Johnson's time in the 100m by first calculating his average speed in the 40-yard dash:

$$v = \frac{\Delta x}{\Delta t} = \frac{\left(40 \text{ yd}\right)}{\left(4.24 \text{ s}\right)} = 8.6 \text{ m/s}$$

Using this speed, we could then calculate his time in the 100m:

$$\Delta t = \frac{\Delta x}{v} = \frac{\left(100 \text{ m}\right)}{\left(8.6 \text{ m/s}^2\right)} = 11.6 \text{ s}$$

That's barely a decent time for a high school student, never mind the fastest man in the NFL. Bolt would blow him away if this were his actual speed. Where did we go wrong? We calculated his speed using his 40 time, but much of the first 40 yards of a sprint is spent accelerating from a low speed. This underestimates Johnson's actual speed, so we can only use it to calculate an upper bound on Johnson's time. If we assume Johnson accelerates uniformly for the whole run, we can calculate his initial acceleration:

$$a_0 = 2 \cdot \frac{\Delta x}{\Delta t^2} = 2 \cdot \frac{\left(40 \text{ yd}\right)}{\left(4.24 \text{ s}^2\right)} = 4.07 \text{ m/s}^2$$

If we then assume he accelerates uniformly throughout the race, we can calculate his new time,

$$\Delta t = \sqrt{\frac{2 \cdot \Delta x}{a_0}} = \sqrt{\frac{2 \cdot (100 \text{ m})}{(4.07 \text{ m/s}^2)}} = 7.01 \text{ s}$$

That's better than even the most unrealistic tricked-out *Madden* player you can make. Because we assumed he was accelerating throughout the whole race, he has a much higher final speed than he would in reality. How much higher? About 64 mph!

Realistically, Johnson would probably reach his top speed somewhere between 10 and 40 yards and then maintain a roughly constant speed throughout the rest of the race. At the very least, he would certainly finish somewhere between these two bounds. Logic would seem to indicate that Bolt, typically regarded as the fastest man in the world, would still win handily. Unfortunately, our simple estimation is not precise enough to say who would win with any certainty. This is an important point. Sometimes we don't have enough background information to calculate definitive answers, and we can only calculate upper and lower bounds. Sadly, we'll have to deal with ESPN pundits endlessly debating who the faster man is.

PHYSICS II:
Baseballs, Footballs, and Other Projectiles

In the previous section, you may have noticed that I only described motion in one dimension, whereas the real world is three-dimensional. I did this mostly for simplicity, but truth be told, there was also a bit of laziness in it. Now is the time to correct this shortcoming. To describe motion in three dimensions, you do the exact same thing you did in one dimension three times, once for left-right motion, once for up-down motion, and once for front-back motion.[43] Physicists call this working with vectors.[44] For example, when you throw a baseball, it travels forward while accelerating downward because of gravity. The forward motion of the ball can be described using the same $x(t)$ equation as the previous section, with the caveat that the acceleration a_0 equals zero because nothing pushes the ball forward once you let it go. Likewise, the same $x(t)$ equation can describe the vertical (i.e., up and down) motion with the caveat that the acceleration a_0 equals the acceleration of gravity g because gravity is the only thing pulling the ball down. At the surface of the Earth, g is nearly constant at 9.8 m/s^2 toward the center of the Earth. Physicists call motion in a gravitational field "projectile" motion.

There are two tricky points about projectile motion. Let's say you drop a baseball straight down. How fast is it moving the instant you let go? The answer is somewhat surprising. It's not! Before you let go, the ball is not moving. Immediately after you let go, it's still not moving. This may be

---

43 When using the same equation to describe motion in two different directions, the variable symbols used in the equation will have multiple meanings. For example, the initial upward velocity of a thrown baseball will not generally be the same as the initial forward velocity. To avoid confusion, you can use subscripts to distinguish variables. For example, to distinguish between the velocity in the x- and y-directions, you can write v_x and v_y, respectively.

44 We will not be working explicitly with *vectors* in this book, but we will use results like the range equation, which are derived from the use of vectors.

counterintuitive, but balls and any other dropped objects start with no initial velocity then accelerate to begin moving. If you have a difficult time believing this, imagine it this way: would you rather have a baseball dropped on your head from a height of 1 millimeter or from the top of the Empire State Building? Clearly, it's better to drop it from as small a height as possible because the ball will be moving with a speed close to the zero velocity it starts with.

The second tricky point is somewhat related. Let's say you throw a ball up in the air. How fast is it moving when it reaches peak height? The answer is the same. It's not moving at the peak of its motion. For the brief instant when the ball is at its peak, it is momentarily still, so its vertical velocity is zero. If you don't buy it, consider the following thought experiment. While the ball is going up, it's constantly slowing down. Eventually it starts going the other way. Mathematically, this is like saying the ball went from having positive velocity to negative velocity. To go continuously from a positive to a negative number, you have to pass through zero, so the velocity at this point must be zero.

With these facts out of the way, we can derive how far a projectile will travel. I'll let you work out the details (it involves using trigonometry), but the horizontal range R of a projectile fired on flat ground is given by

$$R = \frac{v_0^2}{g} \sin(2\theta)$$

where v_0 is the speed at which the projectile leaves, and θ (a Greek letter pronounced "theta") is the angle at which the projectile is fired.[45] The maximum range will occur when θ equals 45°.

Now is probably a good time to mention that I'm going to ignore air friction for simplicity unless otherwise stated. This is not necessarily a good idea. Friction plays a very important role and can alter the numbers significantly, but this is a book about estimation, so leaving it out is only natural. At the end of this chapter, we'll discuss what happens when you include friction. For now, let's just pretend it's small enough that we can ignore it.

---

45 If you've forgotten trigonometry, just remember it's all about right triangles. You may recall learning the pneumonic SOH-CAH-TOA. The SOH tells you that the sine of an angle is defined as the length of the opposite side divided by the hypotenuse (i.e., the longest side.) Similarly, the CAH and TOA tell you that the cosine and tangent of an angle are defined as the length of the adjacent side divided by the hypotenuse and the length of the opposite side divided by the length of the adjacent side, respectively.

You might ask whether or not the v_0 used in this equation refers to the velocity in the vertical or horizontal direction. Let me be picky for a moment and say that this v_0 is actually a speed, not a velocity. When we use the word "velocity," we imply there's a specific direction associated with the motion, but when we say "speed" we mean how fast an object is moving without any regard to direction. You can find the speed by adding the up-down and front-back components of velocity, but be careful! You're adding quantities that are perpendicular to each other, so it's like you're finding the hypotenuse of a right triangle. This can be found by using the Pythagorean theorem,

$$V_0^2 = V_{x,0}^2 + V_{y,0}^2$$

where the $v_{x,0}$ and $v_{y,0}$ are the horizontal and vertical components of the initial velocity.

Blimpitude

The new Cowboy Stadium caused a bit of concern after punters hit the giant scoreboard during warm-ups. At 90 feet above the playing surface, the score-board is only 5 feet above the league-required minimum height. Practically speaking, this is a new problem, but in principle this problem has always existed. Given a strong enough punter, all dome roofs and even planes and blimps are potential obstacles. **How fast would a ball have to be booted to hit a blimp, and what would the hang time be?**

When flying over a sports arena, the Goodyear Blimp typically hovers around 2,000 feet. To find the hang time, we can solve the $x(t)$ equation motion for time t, but we have to double the answer because the ball is going up *and* coming down:

$$t = 2\sqrt{\frac{2\Delta x}{g}} = 2\sqrt{\frac{2 \cdot (2000 \text{ m})}{(9.8 \text{ m/s}^2)}} = 22 \text{ s}$$

Here, Δx is the change in height of the ball, which is equal to the altitude of the blimp. The average hang time for an NFL punt is about one-fifth this result. With 22 seconds of hang time, all 11 members of the kicking team would easily be able to surround the returner as he's catching the ball. Unfortunately, to achieve this hang time, the ball must leave the punter's foot with an extraordinarily large velocity. We can solve for this velocity by multiplying g times the falling time:

$$v = gt = \left(9.8 \text{ m/s}^2\right) \cdot \left(11 \text{ s}\right) = 110 \text{ m/s}$$

At that speed, the football could travel the length of a football field in less than one second.

We could have done the same calculation for an airplane instead of a blimp. With a cruising altitude of 30,000 feet, hitting a plane would require a hang time of 1.5 minutes. Even more impressive, hitting a plane would require a ball to leave the punter's foot with a speed that would break the sound barrier.

Mexican High Jumpers

The 1968 Summer Olympics in Mexico City saw record-setting leaps in the high jump, long jump, and triple jump, not to mention several throwing events. It's been hypothesized that the vat of broken records was due to Mexico City's high altitude, which would create a lower acceleration of gravity.[46] **How much of a difference would the added altitude make on a jump?**

This is a problem best tackled by scaling. As discussed in "Scaling in the Universe" (Part II), the range R of a projectile scales as the inverse of the gravitational acceleration g. From the equation of motion for $x(t)$, you can show the maximum height jumped $h = \Delta x$ will scale inversely as well:

$$h \sim \frac{1}{g} \text{ and } R \sim \frac{1}{g}$$

The gravitational attraction between objects is weaker when the objects are far apart. This fact can be expressed through the scaling relation:

$$g \sim \frac{1}{r^2}$$

where r is the distance between the objects. Combining these two scaling relationships, we see that

$$h \sim r^2 \text{ and } R \sim r^2$$

---

[46] The lighter air may also make a significant difference, but I'm ignoring this for simplicity.

For terrestrial objects, r is the distance from Earth's center to its surface. For the most part, the gravitational acceleration on Earth is roughly constant at about 9.8 m/s². This is because Earth is very spherical, so r is fairly constant, at around 6,378 kilometers. However, in places like Mexico City, the altitude is very high, which can make g change slightly. Mexico City is located 2.2 kilometers above sea level, which would represent a 0.034 percent increase in r. We can write this in scaling terms as

$$r \to r' = 1.00034 \cdot r$$

where r' is the new distance to the center of the Earth. If we assume our high jumpers leave the ground with the same speed and angle they do at sea level, then we can use the scaling relation for h to find the increase in the height:

$$h \to h' \sim r'^2 = \left(1.00034\, r\right)^2 = 1.00069\, r^2 = 1.00069\, h$$

This means the height jumped would increase by about 0.069 percent. Because the range of a projectile scales similarly, the distance in the long jump should also increase by 0.069 percent. Putting this into perspective, you could expect increases of about 1.6 millimeters in the high jump and 6.3 millimeters in the long jump.[47] Jumping records are generally broken by centimeters. Although the decreased gravity would increase the distance jumped, it doesn't have a very large effect.

---

47 This assumes a 2.36-meter high jump and an 8.95-meter long jump.

Dimensional Analysis for Flying Balls

Whenever an announcer reminisces about a mammoth home run, he often recites the now-cliché hyperbole, "He hit a ball that still hasn't landed yet." Although our earthly bodies could never accomplish such a feat, it's still, at least in principle, possible. **How hard would you have to hit a ball so it never lands?**

To begin, I should note that even if a ball is moving very fast, it's still accelerating due to gravity, and that acceleration pulls the ball toward the Earth. "Wouldn't that make it hit the ground?" you ask. That depends. If you drop a ball, it certainly hits the ground. If you hit a ball with a bat, it will start out going straight, but gravity will curve its motion toward the Earth so it eventually hits the ground. If you hit it a little harder, the ball's motion will be a little less curved and it will travel a little farther, but it will still hit the ground. If you hit it hard enough, the amount of curvature in the ball's motion will match the curvature of the Earth, and the ball will not hit the ground.[48] At this point, we can say the ball has entered orbit. This is just like the Moon's orbit, except the ball is very close to the Earth's surface.

This problem is tricky because the ball is moving in a circle.[49] Our problem is to relate the orbital acceleration to the speed the ball must maintain to stay in orbit just a few hundred feet above the ground. We can tackle this using a technique called dimensional analysis. The ball's orbital acceleration is due to gravity (i.e., $g = 9.8$ m/s^2). You might expect the acceleration g for an object moving in a circle at constant speed to be related to the velocity v and the radius of the circle R. Here's one way we can combine these variables to give an equation that has the correct dimensions:

---

[48] "Yes, but wouldn't it come down eventually?" you ask. If it's hit hard enough (and there's no friction), the ball will never return to the Earth. If you doubt this, remember that the Moon has been in orbit for millions of years and still hasn't come crashing into the Earth.

[49] Technically, the orbit is elliptical.

$$g = \frac{v^2}{R}$$

Does this equation make sense physically? Let's say you're running around in a circle. The equation says that g scales as v^2, so a larger speed requires a larger acceleration. This makes sense because the faster you run, the more your velocity changes, which means a bigger acceleration.[50] Put another way, the faster you run, the more your legs have to push to keep you moving in a circle. What about the radius? The equation says that g scales as the inverse of R, so the larger your R, the smaller g has to be. Physically, this makes sense because it's harder to run at full speed in a tight circle than in a larger circle. Put another way, running around in a large circle requires a smaller acceleration because the motion is closer to a straight line. "But," you interject, "you're still moving in a circle! How can it be closer to a straight line?" Good question, Reader. It's never exactly a straight line, but if the circle is *much* larger than you, then at least in your local vicinity it will look like a straight line.[51] At any rate, it turns out we got lucky because dimensional analysis got us exactly the correct equation.

50 Remember, velocity takes into account both speed and direction, so even if you're moving at a constant speed, the change in direction will mean you have an acceleration.

51 This is why people thought the Earth was flat for so long: when you walk forward, you're actually moving along a circular arc, but it seems like you're moving in a straight line because the Earth is much larger than you.

Using the fact the radius of the ball's orbit will be about the same size as Earth's radius $R = 6,400$ kilometers, we can solve for v to find out how hard we have to hit the ball:

$$v = \sqrt{gr} = \sqrt{\left(9.8 \text{ m/s}^2\right) \cdot \left(6400 \text{ km}\right)} = 8000 \text{ m/s.}$$

You would have to hit a ball 8,000 m/s, or roughly 20 times the speed of sound, for it to never hit the ground.

Even more important than our answer, we've added a powerful mathematical tool to our estimation kit. By simply combining the variables you expect your answer to depend on into a form that produces the same dimensions as the thing you're trying to find, you can write approximate solutions to almost any problem. I won't belabor the point too much because we have more fun problems to do, but dimensional analysis is a great tool for estimations. Case in point: G. I. Taylor once estimated the yield of the first atomic bomb after viewing a series of published photographs taken of the event. As physicist David L. Goodstein eloquently wrote, "For the practitioner of the art of dimensional analysis, the nation's deepest secret had been published in *Life* magazine."[52]

52 David L. Goodstein, States of Matter (Englewood Cliffs, NJ: Prentice-Hall, 1975).

PHYSICS III:

Unstoppable Forces

To understand why Clay Matthews easily bowls over running backs, you need to know about forces. Isaac Newton's three laws describe most of what you need to know, but oftentimes these laws are phrased in very confusing ways. Below I've written the laws and provided my own translations, which will hopefully make them easier to understand.

First Law: Objects retain a constant velocity unless acted on by an outside force.

Translation: Things move in a straight line at constant speed . . . unless, of course, they don't.

When you put it that way, it sounds trivial. Surely, there has to be deep meaning here somewhere. After all, when you roll a ball on the floor, it doesn't keep moving at the same speed forever. Eventually it slows down and stops. Newton must have had some deep insight to recognize that if you got rid of all the friction, the ball would keep moving at the same velocity for all eternity.[53] Perhaps, but Newton hasn't bothered to tell us what a force is yet. From the way the law is phrased, one could surmise that a force is something that stops an object from moving in a straight line at constant speed, but it would be silly to define it this way because the logic is circular. Perhaps Newton defines force in the second law.

Second Law: The net force F on an object is equal to the object's mass m times its acceleration a:

$$F = ma$$

Translation: A force is something that stops an object from moving in a straight line at constant speed.

---

53 Physicists call the resistance to changes in velocity "inertia."

Great! Thanks for nothing, Newton. You've just given us a completely circular argument:

Newton: You know there's a force on an object if it accelerates.
You: So what exactly *is* a force?
Newton: It's something that accelerates an object.

That's the logical equivalent of saying Michael Jordan was a great basketball player because Michael Jordan was great at playing basketball.

Fortunately, there is a way out of this logical Möbius strip. Newton's laws are good for making predictions. You know that whenever you throw a baseball, it always accelerates toward the Earth at the same 9.8 m/s^2. You can reproduce this phenomenon however many times you like. Similarly, if you measure a force in one experiment, you're pretty much guaranteed to measure the exact same force in all future experiments, provided you keep all the conditions identical. If you know all the forces, you can predict how everything in the universe will evolve over time. Even better, understanding the universe seems to require only four fundamental types of forces.[54] Newton's first two laws may be circular, but they're pretty darn useful. We may never know exactly what a force *is,* but practically, this doesn't matter. We know what forces *do*, and that's enough to make predictions. At any rate, let's leave the ontological questions to the philosophers and see what Newton has to say in his third law.

Third Law: When object A exerts a force on object B, object A feels a force that is equal in magnitude and opposite in direction to the force exerted on B.

Translation: Whenever I push you, you're also pushing me.

---

[54] These fundamental forces are called gravitational, electromagnetic, strong, and weak. The last two are only noticeable at the subatomic level.

Finally Newton gives us something that's not just circular reasoning.[55] When doing a bench press, you must exert a force up on the barbell. At the same time you're pushing up, you can feel the skin in your hand compressing downward. The weight pushes on you with the same force that you push on it, except it's pushing you down. You might think this downward force is due to gravity, but that's not the downward force we're talking about. Gravity certainly pulls down on the weight, but we're talking about the reaction force and this would be present even if you lifted the weight in outer space. The gravitational force pulling the weight down has its own reaction force, namely, the gravitational force the weight exerts on the Earth.[56]

Newton's third law is a little counterintuitive. After all, when Brock Lesnar punches another fighter in the face, you generally think of the force of the fist on the face rather than the force of the face on the fist, even though both forces are present. When Mike Tyson bit Evander Holyfield's ear, you could truthfully say the ear was pushing on Tyson's teeth with the same force. The only reason Tyson's teeth weren't damaged as much as Holyfield's ear is because teeth are made of a harder material.

Below are some problems that involve force. Before diving in, I should point out that the force of gravity at the surface of the Earth is given by the equation

$$F = mg$$

You can easily see this using Newton's second law and the fact that objects dropped at the surface of the Earth accelerate at a rate $g = 9.8$ m/s^2. More fundamentally, the gravitational force between two objects is given by

$$F = \frac{GMm}{r^2}$$

---

55 This law can be used to derive the conservation of energy and momentum.

56 You might wonder why, if there are equal and opposite forces, the Earth doesn't go flying whenever you jump. This is because the Earth is so much more massive that its acceleration must be very small to produce an equal force.

where M and m are the masses of the two objects that are pulling on each other, r is the separation between the objects, and $G = 6.67 \times 10^{-11}$ N·m²/kg² is a fundamental constant.[57] This force describes the interaction between all masses. Using this formula, you can derive the gravitational force at the Earth's surface by plugging in the Earth's mass and radius for M and r, respectively.

---

[57] The "N" in the unit of gravitational constant stands for "Newton" and is a unit of force. One Newton is equivalent to 1 kg·m/s². Other units of force include pounds (lbs) and dynes (dyn).

Dragoooo!

There's nothing quite like those *Rocky* training montages to get you in the mood to work out. Something about "Eye of the Tiger" and "Hearts on Fire" just gets you pumped to lift heavy weights, but my strength coach, the Muscleless Wonder, noticed something odd about the training scene in *Rocky IV*. At the end of the montage, Rocky tries to look badass by lifting up three people in a cart. There's only one problem. All three people are on the side of the cart with the wheels whereas Sylvester Stallone lifts the side with the handles. **How much weight did Sly actually lift?** [58]

58 In physics-speak, "weight" is technically a force, not a mass, which is why I've included it in this section. For those of us who spend our entire lives on the surface of the Earth, the two terms are mostly interchangeable since weight is proportional to mass via the $F = mg$ equation. For the purposes of this problem, I'm going to solve for mass instead of force since that's what most people are more familiar with.

From the movie, it seems like Rocky lifts up two handles and only slightly tilts back the three people at the opposite end of the cart. The two handles are about the size of 2x4s and they're around 7 feet long, giving a total volume of about 22,000 cubic centimeters. Even if the cart were made of a heavy wood like ebony, which has a density $\rho = 1,100$ kg/m^3, the total mass lifted would only be about [59]

$$m = \rho V = \left(1100 \text{ kg/m}^3\right) \cdot \left(22,000 \text{ cm}^2\right) = 24 \text{ kg.}$$

A lighter wood like Douglas fir would be about half this weight, meaning Rocky only pressed about 12 kilograms (~27 pounds). Put another way, while steroid-pumping Drago is pressing 185 kilograms (~405 pounds), Rocky was struggling to pick up two kitty cats. Perhaps Sly really did need those 48 vials of HGH he got caught smuggling into Australia.

---

[59] The symbol "ρ" is a Greek letter pronounced "rho." It's used to represent density, which we'll discuss in greater detail when we get to the section titled "Physics VII: Under Pressure (Buoyancy and Temperature)."

Send in the Clowns

I often wonder about the origin of things.[60] Take bull riding. What idiot thought it would be a good idea to jump on the back of a horned, 2,000-pound monster with a bad attitude and a severe dislike of the low frequency end of the visible spectrum?[61] I can't help but think it was some old cowboys' version of *Jackass*. Still, you've got to respect these eight-second wonders for having enough ~~suicidal tendencies~~ bravery to even attempt this sport. **With how much force must a bull buck to guarantee the rider will fall?**

Strictly speaking, a bull's bucking is not what causes the rider to fall off; it's the rider's own inertia and internal forces that lead to the inevitable dismount. When the bull bucks up, the rider's body compresses and accelerates with the bull. After the buck, there are two possible mechanisms that create space between the rider and the bull. First, the bull could buck downward. According to Newton, a body moving at constant velocity will keep moving at constant velocity unless something forces it off its path. If the body in question is that of the rider, his inertia will want to keep him moving upward even as the bull has bucked itself downward. If this wasn't enough, there's still a small effect due to forces inside the rider's body. As the bull bucks up, the rider's body compresses slightly. Much like a spring, the rider's body will want to expand to its normal state. As the body expands, it pushes against the bull and may propel the rider upward. Some combination of these two effects lead to a projectile rider. The only thing keeping the rider connected is the gripping force of one hand on a flat braided rope attached to the bull.

According to Wikipedia, the record for a one-handed deadlift is 330 kilograms (~727 pounds) by Hermann Görner. Because this is a world record, we can reasonably treat this as an upper bound for grip strength. By using the

---

60 On a bovine-related origins note, how did people start drinking milk? Who was the first human to look at a cow udder slowly dripping a strange cloudy fluid and think, "Mmmm! That looks tasty!" What was he thinking? Perhaps more entertainingly, what was that first cow thinking?

61 In 2007, MythBusters actually busted the myth that bulls hate the color red. Perhaps I should stop getting my animal facts from Warner Bros. cartoons.

gravitational force equation, we can compute the total force the hand must exert on the weight:

$$F = mg = (330 \text{ kg}) \cdot (9.8 \text{ m/s}^2) = 3200 \text{ N}.$$

Assuming an 80 kilogram (~175 pound) rider, we can use Newton's second law to compute the maximum acceleration the rider can withstand with this force:[62]

$$a = \frac{F}{m} = \frac{(3200 \text{ N})}{(80 \text{ kg})} = 40 \text{ m/s}^2$$

This is the maximum a bull must accelerate its body to knock the rider off. To accelerate its whole body, a 2,000-pound bull must apply a force of

$$F = ma = (2000 \text{ lbs}) \cdot (40 \text{ m/s}^2) = 36,000 \text{ N}.$$

That's 36,000 Newtons, or a force roughly equivalent to four times the bull's weight.

There's a problem with this estimate. I've assumed the bull is completely rigid here, which, of course, is not the case. Anyone who's seen bull riding knows the bull wildly contorts its body in an attempt to fling the rider. To accomplish its goal, the bull doesn't need to accelerate all of its mass at once, just the part of its body that touches the rider, so the force needed can be much smaller. As such, our result is the maximum force the bull must apply to its body, but it can probably buck the rider off with appreciably less.

62 We're only interested in the motion of the rider relative to the bull. For this reason, we can ignore the force of gravity at least for the brief instance when the bull and rider are in midair because gravity will accelerate the two at the same rate.

Moon Shot

In 1989, Culture Brain released *Baseball Simulator 1.000* for the Nintendo Entertainment System. The game had the unique feature that allowed you to choose the type of stadium you wanted to play in. One of the options was "Space." **How much farther would a ball travel in a space stadium than in an Earth stadium?**

"Space" is a pretty poorly defined term here. If we're talking about the Moon, then gravity is about one-sixth as strong as it is on the Earth. If we're talking about Jupiter, gravity is about 2.5 times stronger than it is on the surface of the Earth. However, the space stadium in the game appears to be a space station floating in free space. What would gravity be there?

The International Space Station is 51 meters long, 109 meters wide and 20 meters tall with a mass of about 417,000 kilograms. Considering the dimensions of a typical baseball stadium, it seems reasonable that a space stadium would need to be between 10 and 1,000 times more massive than the Space Station. Using these bounds, we can estimate the total mass of the stadium to be equivalent to 100 international space stations.

Upper Bound: 1,000 International Space Stations
Assumed Average: 100 International Space Stations
Lower Bound: 10 International Space Stations

The center of mass of the space station might be about 20 meters below the playing surface.[63] From these assumptions, we can calculate the gravitational acceleration g' in the stadium:

$$g' = \frac{F}{m} = \frac{GM}{r^2} = \frac{\left(6.67 \times 10^{-11}\ \text{N} \cdot \text{m}^2 / \text{kg}^2\right) \cdot \left(4.17 \times 10^7\ \text{kg}\right)}{\left(20\ \text{m}\right)^2} = 7.5 \times 10^{-6}\ \text{m/s}^2$$

That's about 0.000076 percent the gravitational acceleration of the Earth. Using scaling, we can find the range R' of a baseball hit in a space stadium:

$$R \rightarrow R' \sim \frac{1}{g'} = \frac{1}{\left(0.00000076\ g\right)} = 1.3 \times 10^6\ R$$

A baseball would travel 1.3 million times farther in a space stadium. This means a 400-foot home run would travel 100,000 miles! Needless to say, my *Baseball Simulator* home runs never went that far.

---

63 I'm ignoring the fact that a space stadium has a nonspherical geometry. In principle, one should integrate (i.e., add up) all the little chunks of matter to compute g. The actual value for g will be somewhat smaller than the one computed here.

Remember to Breathe When You Lift

In weightlifting, I don't think sudden, uncontrolled urination should automatically disqualify you.

—Jack Handey, in *Deep Thoughts*

Jack Handey might have been happy to know that during my extensive research conducted over the last two minutes, I have found no such provision in the International Weightlifting Federation (IWF) rules. Moreover, it's unclear why the IWF would support any such rule, as uncontrolled urination would seem to provide no discernable unfair advantage for lifters in a competition. In contrast to urination, one might expect that uncontrolled flatulence may provide a slight advantage because the gas would help propel the lifter upward. **How much extra weight could a lifter with uncontrollable flatulence squat?**

Farts come in a variety of shapes and sizes.[64] Assuming they have a volume of about ten cubic centimeters and the same density as air (about 1.2 kg/m^3), each fart will weigh about ten milligrams. Let's assume they come out at a speed of 1.0 m/s and last only 1.0 seconds. Using this data and dimensional analysis, we can estimate the total force:

$$F \approx \frac{mv}{t} = \frac{(10 \text{ mg}) \cdot (1.0 \text{ m/s})}{(1.0 \text{ s})} = 0.01 \text{ N}$$

Continuous flatulence would produce a 0.01-Newton reaction force upward, which is equivalent to lifting an extra 0.002 pounds.

---

64 Not to mention flavors.

Atlas Shrugs

Give me a place to stand, and a lever long enough, and I will move the world.

—Archimedes

I had shoulder surgery a few years ago. During rehab an athletic trainer had me doing lat pulldowns to strengthen the muscles. For those unfamiliar with this exercise, it's basically like doing a chin up, but rather than lifting your chin to the bar, you pull the bar down to your chin. Attached to the bar is a cable that passes through a loop above your head before coming down and connecting to some adjustable weights.

All was going better than expected. Fifty pounds? No problem. One hundred pounds? Piece of cake. Two hundred pounds? Barely breaking a sweat. Three hundred pounds? At this point, the skeptic in me awoke. How could the

180 pounds of me lift the 300 pounds of weight on the opposite end of the cable? Shouldn't my body be going up instead of the weight? Much to the dismay of my ego, I looked inside the Nautilus machine and found not one but *two* pulleys attached to the weight.

When you pull on the end of a cable, each little cable molecule pulls on the cable molecule next to it. This pulling force is called tension. When pulling up on a cable attached to a weight, this tension force is transferred from one molecule to the next until it reaches the molecules in the weight. If the tension force is stronger than the force of gravity, the weight will rise up. A strange thing happens when you attach a pulley to the weight. Because the pulley has one end of the cable going in and one end of the cable coming out, it will experience twice as much tension force pulling up. By attaching a pulley to a weight, you need half as much force to lift the weight. In this way, pulleys provide a mechanical advantage.[65] In principle, you can lift an object of any weight provided you have enough pulleys. **How many pulleys would you need to lift the Earth?** [66]

In 1798 Henry Cavendish measured the mass of the Earth. The currently accepted value is 5.97×10^{24} kilograms. A reasonably strong person might be able to lift 45 kilograms (~100 pounds), which means s/he needs 1.3×10^{23} times as much force to lift an object the size of the Earth. After adding one pulley, there are two cables pulling up on the weight, so you get twice as much force. After adding two pulleys, you get four cables pulling up on the weight so you get four times as much force. After adding three pulleys . . . it's easy to see the pattern. If you add N pulleys to the weight, you will get $2N$ times as much force. We can solve for the number of pulleys to get $N = 6.5 \times 10^{22}$. Given a six-inch-wide pulley, you would need enough pulleys to cover the Earth 3 million times.

---

65 You might think you're getting something for nothing with this whole pulley business, but there's always a tradeoff. In this case, the tradeoff is that for every pulley you add, you have to pull the rope farther to make the weight travel the same distance. For example, if you're using a single pulley to lift an object one meter off the ground, you need to pull the rope two meters. If you want to do the same thing but with two pulleys attached to the object, then you have to pull the rope four meters.

66 I should clarify. Anyone doing a handstand looks like he's lifting the Earth, but this is not what I mean. The phrase "lift the Earth" doesn't make logical sense. The Earth doesn't pull on itself, so it can't have a weight per se. I should have asked, "How many pulleys would it take to lift an object that has the same mass as the Earth?," but this sounds less dramatic.

A Clean Sweep

In an episode of *The Simpsons* titled "Boy Meets Curl," Homer and Marge take up curling. After a particularly bad delivery by Homer, Marge injures her arm sweeping the ice while trying to correct the stone's motion. In curling, the sweeper can guide the motion of the stone by reducing the friction on different parts of the ice. In principle, if you sweep enough ice away, you can start a stone even after it's stopped. **How much of an incline would you need to sweep out to start a stopped stone?**

A curling stone weighs about 20 kilograms (~40 pounds). To start it moving from rest, you need to push harder than the frictional force preventing its movement. To figure out the angle needed, we first need to find the force required to initiate its movement. Imagine a curling stone placed on a table made of ice. Let's say you tie a string to the stone and attach a weight to the opposite end of the string. You let the weight hang over the side of the ice table. How much weight would you need to add before the stone starts to move? As a rough estimate, we might say about one kilogram if the ice surface is very slippery.[67] This corresponds to a 10 Newton force.

In order to start the stone moving from rest, you need to scrape away enough ice that gravity will overcome the 10 Newton frictional force and pull the stone down the slippery slope. In free fall, the force of gravity is given by the equation $F = mg$, but our stone is not in free fall. Because it's sliding down a slope at an angle, it won't feel the full effect of gravity. If the slope of the ice were 0° (i.e., completely horizontal), gravity wouldn't be able to move the stone at all. In this case, the gravitational force is effectively zero. At the other extreme, you have a slope of 90° (i.e., completely vertical). This would be identical to free fall because there's nothing for the stone to slide down, so

---

67 If this seems unreasonable, consider your bounds. Intuitively, 10 kilograms seems like too much, but 0.1 kilogram seems too small. If you're well versed in physics, you may want to try computing the same result using the coefficient of static friction between the ice sheet and the curling stone. According to hypertextbook.com, the frictional coefficient for ice and a curling stone is about 0.0168.

the force would just be *mg*. To relate these two extremes, we can use our old trigonometric friend, the "sin θ" function, which is equal to 0 when $\theta = 0$ and 1 when $\theta = 1$. Using this, we can write the effective force of gravity as [68]

$$F = mg\sin\theta$$

From this, we can solve for:

$$\theta = \sin^{-1}\left(\frac{F}{mg}\right) = \sin^{-1}\left(\frac{10\ \text{N}}{(20\ \text{kg})\cdot(9.8\ \text{m/s}^2)}\right) = 3°$$

A 3° angle, doesn't sound like much, but if the stone stops halfway down the ice, the sweeper would need to sweep away about three cubic meters worth of ice for the stone to reach the circle. It's no wonder Marge injured herself.

---

[68] Picky readers will note that I haven't actually proven that F = mg sin θ is correct except for the cases when θ is either 0 or 90°. It's a fair critique, but I'd rather keep this simple even if it is a little hand waving. Those readers who have taken physics will recognize this as a vector projection. Physicists use vectors to deal with three-dimensional forces, among other things. When you do the vector projection, you obtain the same result used here.

It's Like Steroids on Steroids

Be invisible. Because I'm a pervert, dude. I'd be in every girls' bathroom, locker rooms. All of it.

—Pitcher Barry Zito in an interview with ESPN's "Page 2" when asked if he would prefer having strength of 1,000 men, the ability to fly, or invisibility

Barry Zito displayed some refreshing honesty in his interview, but I suspect most professional athletes would choose the strength of 1,000 men, given how much of an advantage it would provide on the ball field. Invisibility and flight may only exist outside the realm of reality, but with the advent of steroids, HGH, and other performance-enhancing drugs (PEDs), unnatural strength gains are quickly becoming the norm in many professional sports. Given the clear advantage strength provides in athletic competition, it's ironic that so many baseball players from the steroid era deny that PEDs help you hit a baseball. **How would a modest 10 percent increase in strength help with hitting a baseball?**

The term "strength" here is a little vague. I interpret a "10 percent increase in strength" as a 10 percent increase in the force a player's muscles can supply to the bat. Given that the physics of hitting a baseball is fairly complicated, it's difficult to get precise numbers, but we can get a rough idea of the impact increased strength would have by considering three examples: reaction time, bat speed, and fly ball distance. We can tackle these problems quickly and effectively using scaling arguments.

Reaction Time

When a player swings, the bat accelerates through the strike zone. Regardless of the player's strength, the distance swung remains constant, even though the bat speed and the acceleration may change. From the equations of motion for $x(t)$, the displacement x of an object is given by

$$\Delta x = \frac{1}{2} at^2$$

We can see that the time t it takes for the bat to reach the strike zone scales with the acceleration as

$$t \sim a^{-1/2}$$

Because force and acceleration are directly proportional, a 10 percent increase in the force will generate a 10 percent increase in the acceleration,

$$a \to a' = 1.1\,a$$

which leads to a decreased t':

$$t' \sim a'^{-1/2} = \left(1.10\,a\right)^{-1/2} = 0.95\,a^{-1/2} \sim 0.95\,t$$

That's 5 percent less time required. Assuming a hitter accelerates the bat in 0.4 seconds, the extra speed will give him an extra 0.02 seconds to determine what type of pitch has been thrown. That may not sound like much, but it would be noticeable to a Major League hitter. It's the difference between hitting a 97-mph fastball and a 92-mph fastball.

Bat Speed

Extra reaction time isn't the only benefit of increased strength. Using the equation of motion for velocity, one can see the change in velocity v is given by

$$\Delta v = at$$

By combining this with the scaling relation for the time derived above, we can see that the bat speed through the strike zone scales with acceleration as

$$v \sim a^{1/2}$$

The increase in bat speed is given by

$$v' \sim a'^{1/2} = \left(1.10\ a\right)^{1/2} = 1.05\ a^{3/2} = 1.05\ v$$

which represents a 5 percent increase in bat speed. In a 2006 experiment at Washington University, Albert Pujols swung a 31.5-ounce bat 86.99 mph. A 5 percent increase in speed would increase his bat speed to 91.34 mph. You don't need Roger Clemens's third ear to see—er—hear that extra bat speed is an advantage. For example, a ball that an infielder could previously have fielded cleanly is now very much out of range because it's traveling too fast for him to reach.

Fly Ball Distance

The last effect, fly ball distance, is almost certainly the most noticeable given the rash of home runs hit during the steroid era. For simplicity, I'll assume the speed with which the ball leaves the bat increases by the same 5 percent that bat speed does.[69] Using this assumption and the fact that the range of a projectile scales as

$$R \sim v^2$$

we can show that the increased range R' will be given by

$$R' \sim v'^2 = (1.05\ v)^2 = 1.10\ v^2 \sim 1.10\ R$$

That's a 10 percent increase in the range of a fly ball. Put into perspective, what used to be a 365-foot fly out is now an over 400-foot home run. To be fair, we've neglected air resistance, so the results aren't quite that dramatic, but even with friction, you're bound to see a lot more home runs.[70]

---

69 This is actually a fairly difficult problem if you want to get exact numbers, but you can get reasonable estimates by using conservation on momentum and energy. For now, I'll just take this as an order of magnitude estimate.

70 If you include air resistance, it would take a 370-foot fly out to equal a 400-foot home run.

PHYSICS IV:
Conserving Energy

Energy is a strange beast. Much like force, we don't know what it is, but we do know how to calculate it. When you push on an object, you give it some energy. The amount of energy given to the object can be calculated by multiplying the force applied F times the distance x that the object is pushed:[71]

$$E = F \cdot x$$

The weird thing about energy is that the total amount always seems to stay the same.[72] Why then, pray tell, do athletes sometimes "run out of energy" at the end of a game? Although the total energy of the universe is constant, energy can change from one form to another in a variety of ways. For example, a marathoner converts a large fraction of her body's chemical energy into the kinetic energy of motion:

$$KE = \frac{1}{2}mv^2$$

where m is the mass of the moving body and v is the velocity. In contrast, a weightlifter converts his chemical energy into both kinetic and gravitational potential energy:

---

71 For simplicity, I've assumed the force is both constant and pushing in the direction of motion, but this quantity can be calculated more generally by using vectors and integration. Physicists call this quantity the "work" done on the object. It and all other types of energy can be expressed in units of J (pronounced "joule" after physicist James Prescott Joule). Expressed in other units, one J is equivalent to one $kg \cdot m^2/s^2$. Other common energy units include kilowatt-hours (kWh), British thermal units (Btu), electron volts (eV), and calories (cal). When referring to food, the unit "calorie," sometimes called "food calorie," actually equals 1,000 times the "calorie" unit that appears in physics texts. Because it is more commonly associated with food, I will use the word "calorie" to mean a food calorie.

72 I won't go into detail, but this result can be derived from Newton's third law.

$$GPE = mgh$$

where g is the acceleration of gravity defined previously, m is the mass of the weight, and h is the height that the weight is lifted. If you add up all the chemical energy an athlete starts with, the result will be equal to all the energy he is left with plus any energy he transferred to other objects. The rate at which energy gets converted from one form to another is called power P, and it can be approximated using the equation

$$P = \frac{\Delta E}{\Delta t}$$

where ΔE is the amount of energy that changes form and Δt is the time that passes during the transfer.

In a very general sense, machines are devices for converting energy from one form to another. In this regard, Albert Pujols's nickname "The Machine" is fitting because all athletes convert chemical energy into motion of some form. However, it's important to note that all machines lose a little bit of energy in the conversion process. Although the total energy is always conserved, some of it inevitably goes into pushing air and other molecules that move around randomly. Because the temperature of an object is correlated to the random motion of molecules inside, the end result of this energy loss is that some pieces of matter get a little bit hotter. This is not particularly productive, as that energy could have been used for something more practical. A machine engine that transfers energy from one form to another without losing much in the process is said to be efficient. It can be said that Albert Pujols has a relatively efficient swing because much of his chemical energy is transferred to the flight of the ball.

The problems in this section deal with energy, power, and efficiency in the world of sports and exercise. As before, we'll not be dealing explicitly with friction unless otherwise stated.

Breakfast of Champions

There are many who argue that competitive eating is not a sport. "True athletic competitions," they say, "require speed, strength, and power." Well, it certainly requires speed to gulp down a dozen hot dogs in under a minute and strength not to puke it all onto your opponent's face. But what about power? **What is Takeru Kobayashi's power consumption in a hot dog eating contest?**

In 2007, Kobayashi placed second by eating 63 hot dogs in 12 minutes. One hot dog and bun have roughly 250 calories, so 63 hot dogs and buns must have about 16,000 calories. We generally think of a "calorie" as a measure of how much food we've eaten, but it's actually an energy unit. Power is a measure of how quickly energy flows from one form to another. To calculate Kobayashi's power, we can use the power equation

$$P = \frac{\Delta E}{\Delta t} = \frac{\left(16,000 \text{ Calories}\right)}{\left(12 \text{ minutes}\right)} = 9.0 \times 10^4 \text{ W}$$

Kobayashi's power consumption is about 90,000 Watts. To the naysayers out there who claim that competitive eating is not a sport, I say this: A 95-kilogram Tour de France cyclist accelerating to 25 mph in 5 seconds still outputs only one-third the power that Kobayashi consumes in a competition!

The Phelps Diet

Shortly after winning eight gold medals at the Beijing Olympics, Michael Phelps appeared as a guest on *Saturday Night Live*. In one bit, Phelps touts his 12,000-calorie-a-day diet plan in an infomercial-style skit while several of the actors grow morbidly obese trying to follow it. A warning appears at the bottom of the screen stating, "caloric intake based on 4,000 laps a day at world-record pace." Because of their huge energy output, professional athletes must consume many more calories than the average human. **Based on the 12,000-calorie figure, what is Michael Phelps's average power output during practice?**

A typical human consumes about 2,000 calories per day, which is equivalent to 100 watts.[73]

At 12,000 calories, Phelps consumes 10,000 calories more than a normal person. The main difference between Phelps and a normal human is, of course, the 4,000 laps per day at world-record pace. Presumably, this extra work accounts for the added caloric expenditure.[74] Assuming a five-hour workout, Phelps's power output would have to be

$$P = \frac{\Delta E}{\Delta t} = \frac{(10,000 \text{ Calories})}{(5 \text{ hours})} = 2,300 \text{ W}$$

During a workout Michael Phelps is about 23 times more powerful than you.

---

73 That's right, the human body uses energy at about the same rate as a light bulb.

74 I need to make a disclaimer. I'm a physicist, not a biologist, so I reserve the right to butcher any and all biology statements made in this book. Like many physicists, I tend to assume the complexity of biological systems can be reduced by making a series of broad approximations. Sadly, this doesn't always work. In this example, I'm assuming all the extra energy Phelps consumes is used during his workouts. It's not at all clear this is the case. Professional athletes generally have a higher resting metabolic rate than a normal person. For example, extra energy may be required to repair tissue broken during practice. If this is the case, then Phelps's power output during the workout will almost certainly be smaller than the one calculated here.

Weight Loss Episode I: Stair Master

Since it's clear the Michael Phelps diet plan is not going to help your weight loss goals, I thought I might recommend a few surefire tips to help shed those extra pounds. To aid in this venture, I've elicited the help of my strength coach friend, the Muscleless Wonder. One day at the gym, MW couldn't help noticing a strange bit of irony in the fact that people pay hundreds of dollars each year to pretend they're going up and down something that can be found in any two-story home. Although Stairmasters are a favorite exercise among many casual exercise enthusiasts, the real thing would be cheaper and just as effective. **How many stairs would you need to climb to be guaranteed to lose ten pounds of fat?**

Moving up in a gravitational field requires energy. In the case of stair climbing, the chemicals in the body supply that energy. In high school biology, you probably learned that ATP is the molecule that stores this energy, but once ATP runs out, you start burning fat. For simplicity, I'm going to ignore the role of ATP and just assume you're burning fat.[75] Every gram of fat stores about nine calories of energy. From this, we can calculate how much energy is stored in ten pounds of fat:

$$\text{total energy} = (\text{mass of fat}) \cdot (\text{energy per unit mass})$$
$$= (10.0 \text{ lbs}) \cdot (9 \text{ calories/gram})$$
$$= 1.7 \times 10^8 \text{ J.}$$

---

[75] There's some sketchy biology going on here, and I don't blame you if you find faults with the assumptions. I assure you, the answer to this problem will be so absurd that neglecting the ADP/ATP cycle will be the least of your worries.

To burn this fat, you need to convert its energy E into kinetic energy, which will in turn get converted to gravitational potential energy GPE. If we assume 100 percent efficiency, then we can calculate how high you'd have to climb using the gravitational potential energy equation:

$$GPE = mgh$$

Assuming you finish with a mass of 90 kilograms (~200 pounds),[76] you can solve for the height h you'd have to climb.

$$h = \frac{E}{mg} = \frac{\left(1.7 \times 10^8 \text{ J}\right)}{(90 \text{ kg}) \cdot \left(9.8 \text{ m/s}^2\right)} = 190 \text{ km}$$

That's 190 kilometers, or roughly the equivalent of climbing Mount Everest 22 times. Assuming one foot per stair, you'd need to climb 600,000 stairs. At a rate of one stair per second, you'd need to climb for over a week straight to lose ten pounds of fat.

To be fair, we've assumed our bodies are 100 percent efficient, when in actuality our skeletal muscles are only about 20 percent efficient.[77] Taking into account this inefficiency, one would only have to climb 38 kilometers worth of stairs, which would take roughly 1.4 days.

---

76 For simplicity, I'm ignoring the fact that weight is changing continuously. This will be a small (< 10 percent) contribution to the overall energy and will not affect the order of magnitude estimate.

77 I'm citing this figure from John Eric Goff's very readable *Gold Medal Physics: The Science of Sports*. (Baltimore, MD: Johns Hopkins University Press, 2010). One can easily see this value is only approximate by considering isometric exercises, which do not involve movement. Despite the lack of motion, these exercises can consume quite a lot of chemical energy, which would mean they have zero efficiency.

Weight Loss Episode II: The Bench

Me: *Hey, Muscleless Wonder, is it common to make noises when you're working out?*

MW: *It happens.*

Me: *Yeah . . . but . . . there's some guy over there doing bench presses and making sex noises.*

MW: *Maybe he really likes the bench.*

I suppose it's possible to like benching too much, but if you do, then this workout's for you. **How much would you have to bench in a workout to lose ten pounds?**

Unlike the previous problem, we're solving here for the weight lifted rather than the height climbed. In this example, the height is measured from the bottom position of the bar at your chest to the top position of the bar when your arms are fully extended. This will be about two feet.

From the previous problem, we know ten pounds of fat is equivalent to 1.7×10^8 J of energy. Now, we could use this result and simply solve the gravitational potential energy equation for the mass m required to lose ten pounds on a single rep of the bench press, but doing so would not pay homage to the holy three sets of ten repetitions that every workout seems to require. I'm not sure where this golden rule came from, but it's the standard for almost every workout. For some reason, three-by-ten is treated with the same sort of reverence that mathematicians give numbers like e and π, and far be it for me to break with exercise tradition.

We can view the three sets of ten repetitions as being equivalent to lifting the same weight 30 times higher, or roughly 60 feet. Solving for the mass m, we get

$$m = \frac{E}{gh} = \frac{\left(1.7 \times 10^8 \text{ J}\right)}{\left(9.8 \text{ m/s}^2\right) \cdot \left(60 \text{ ft}\right)} = 95{,}000 \text{ kg}$$

That's 95,000 kilograms, or roughly the equivalent of lifting a blue whale for three sets of ten. If we take into account efficiency, it's still like lifting a humpback whale.

Weight Loss Episode III: Beer Curls

The beer gut gets a bad reputation. After all, that beer doesn't get to your mouth by itself. Somebody has to do the heavy lifting, and that lifting is going to burn calories. **How much weight do you lose from all the beer curls you do in a lifetime?**

Let's say you're a bit of a lush and you drink three beers a day, every day for 60 years. Each beer you drink must be lifted to the mouth about ten times. From this, we can estimate the total number of beer curls in one lifetime:

$$\text{\# of curls} = (\text{\# of beers per day}) \cdot (\text{\# of curls per beer}) \cdot (\text{\# of years})$$
$$= (2 \text{ beers / day}) \cdot (10 \text{ curls / beer}) \cdot (60 \text{ years})$$
$$= 660{,}000 \text{ beer curls.}$$

A beer curl will raise the can about 1.5 feet. A standard beer can weighs 12 ounces, but on average it's only half full, so I'll assume a mass of 6 ounces. From this and the number of beer curls N, we can estimate the total energy used to lift beer:

$$E = Nmgh = (660,000) \cdot (6 \text{ oz}) \cdot (9.8 \text{ m/s}^2) \cdot (1.5 \text{ ft}) = 500,000 \text{ J}$$

That's 500,000 J. Since there are 9 calories per gram in fat, all the beer curls in your lifetime would burn only 13 grams (~0.5 ounces) of fat. Even if you take into account the efficiency of the muscle, that's only 65 grams (~2.5 ounces). In contrast, a light beer contains about 100 calories per can. Given the above assumptions, the calories from the amount of beer drunk would be equivalent to 730 kilograms (~1,600 pounds) of fat added over the course of a lifetime.

Weight Loss Episode IV: Cosmo Calories

We come now to our final episode of weight-loss physics. Skeptical of the "sex burns hundreds of calories" claims gracing the covers of women's magazines in store checkout lines, the Muscleless Wonder once asked if there was any way to calculate the actual number. It's not clear where these figures come from, but sex can mean a lot of different things, so they probably have a large error range. If you're planning on getting to third base, here's a handy estimation that will help you figure out how many calories you'll burn in the process. **How many calories does a typical sexual encounter burn?**

First off, it should be stated that sex, like any other physical activity, burns calories. This fact is not in dispute, but there is a question of magnitude. As with running, how much you burn depends greatly on whether you're doing a marathon or trying to break Usain Bolt's speed record. Furthermore, the rate of energy consumption is highly variable because the participants' involvement can range from playing every down to treating it like a spectator sport. Assuming you're not a passive observer during the event, your body is probably undergoing a certain amount of—er—undulations. Although the mechanics is different for males and females, the overall motion involves a certain amount of pelvic thrusting. Because the pelvis might make up one-fifth a human's height, it seems logical that this should comprise about one-fifth a human's mass or roughly $m = 15$ kilograms. Each forward thrust might last $t = 0.5$ second and displace the pelvis about $x = 10$ centimeters forward. Using dimensional analysis, we can estimate the power consumption:

$$P \approx \frac{mx^2}{t^3} = \frac{(15 \text{ kg}) \cdot (10 \text{ cm})^2}{(0.5 \text{ s})^3} = 1.2 \text{ W}$$

At a rate of 1.2 watts, sex would burn about one calorie per hour. Taking into account muscle efficiency would bump this up to about five calories per hour.

Given that our result is only expected to be accurate to within an order of magnitude, burning 100 calories during an hour-long romp in the sack seems possible, but this number's a little on the high side. More importantly, it's misleading to cite this number without providing any sort of error range. Earlier, I mentioned how physicists are more concerned with functions than they are with absolute numbers. This is true even when they're talking about sex numbers. Notice that the rate of energy consumption scales with thrusting time as t^3. If a person thrusts twice as fast, the power consumption increases by a factor of eight, (i.e., instead of burning 5 calories, you'd be burning 40 calories.) Similarly, cutting the thrust speed in half will reduce the calories burned by a factor of eight. By altering the values of our variables in this way, we get a good idea what kind of error range our final answer will have. This procedure, sometimes called assumption analysis, is very useful for determining how reliable your estimate is. In our case, it shows that the amount of calories burned greatly depends how vigorous the sex is.

Vault-y-more

Pole vault is an impressive sport. You run, jab a stick in the ground, and go flying 20 feet in the air. It's magnificent, really. And the physics, at least from a conservation of energy standpoint, is fairly straightforward.[78] If you run fast enough, you can, in principle, vault over anything. **How fast would you have to run to vault the Gateway Arch in Saint Louis?**

---

[78] There is rich physics in the bending of the pole, but I'm ignoring this for simplicity.

At 192 meters (~630 feet), the Gateway Arch is the tallest manmade monument in the United States. To pole vault to the top, an athlete must convert his/her kinetic energy,

$$KE = \frac{1}{2}mv^2$$

into gravitational potential energy:

$$GPE = mgh$$

Setting these equal to each other, we can solve for the velocity the pole vaulter needs:

$$v = \sqrt{2gh} = \sqrt{2 \cdot (9.8 \text{ m/s}^2) \cdot (192 \text{ m})} = 61 \text{ m/s}$$

An athlete would have to run at least 61 m/s, or roughly 140 mph, to pole vault the Gateway Arch in Saint Louis.

We've neglected something. How much would the pole for this vault have to weigh? By my estimate, the pole would have to weigh over 200 kilograms (~440 pounds). This brings up an important point. Sometimes after estimating, you realize you made an assumption that was sketchy at best. In this case, we assumed the mass of the pole was negligible, which is mostly true for a normal pole vault but not at all true if you're trying to vault the Gateway Arch. To solve this problem rigorously, we really need to take into account the kinetic and potential energy changes in the pole as well. If you're up to the challenge, see if you can solve this problem with the mass of the pole included.

PHYSICS V:

Momentum Shifts

Energy is not the only quantity the universe likes to conserve. A second conserved quantity is the momentum p, which is equal to the mass m times the velocity v:

$$p = mv$$

When two bodies collide, the total momentum before is always the same as the total momentum after. If football defensive lineman Vincent Wilfork and racehorse jockey John Velazquez ran at full tilt toward each other, it's near certainty that Velazquez would get smashed backward into a bloody pulp. The 110-pound jockey would need to run three times the speed of the 330-pound Wilfork just for a stalemate. Even if Wilfork runs a glacial six-second 40-yard dash time, there's no way Velazquez could knock him back because it would require the jockey to run a 40-yard dash in two seconds.[79]

In addition to energy and linear momentum, there's a third conserved quantity called angular momentum. Angular momentum is similar to momentum except it describes rotations. Any object that rotates (e.g., a spiraling football, a spinning figure skater, a curveball, etc.) has angular momentum. The angular momentum L of a single particle can be calculated using the equation

$$L = mvr$$

where m is the mass of the particle, v is the linear velocity at which it's moving, and r is distance between the particle and the axis of rotation.

Many of the other physical quantities discussed so far have an analogous

---

79 Even the NFL's fastest receivers don't run much below a 4.2-second 40-yard dash.

rotational version. For example, distance traveled is analogous to the angle rotated, and velocity v is analogous to the angular velocity ω (a Greek letter pronounced "omega"):

$$\omega = \frac{v}{r}$$

Angular velocity is typically measured in units of Hz = 1/s (pronounced "hertz" after physicist Heinrich Hertz) or rotations per minute (rpm). For example, a DVD player spins at about 6 Hz or, equivalently, 350 rpm.

Below are a couple examples illustrating momentum conservation in sports.

Spin City

With the exception of an occasional Will Ferrell flick, Americans only seem to care about figure skating once every four years. But you don't need Jon Heder's nutsack nestled up against your chin to appreciate the brilliant physics found in the rotational motion of figure skaters. **How much does a figure skater's angular velocity increase when she pulls her arms in?**

We can compute the angular momentum L of each atom in the skater's body using the formula

$$L = mvr$$

where m is the mass of the atom, v is the velocity of the atom, and r and the distance from the axis of rotation. Because the linear velocity can be expressed in terms of the angular velocity via the equation

$$v = \omega r$$

we can rewrite the angular momentum as

$$L = mr^2\omega$$

If the skater keeps her body rigid, ω will be roughly the same for every atom. When she pulls her arms in, r decreases so the angular velocity ω must increase if angular momentum is to be conserved. Assuming the skater's body is uniformly dense, we can write the scaling relationship between the angular velocity and average radial distance as

$$\omega \sim r^{-2}$$

If we assume the average distance r between the atoms in the skater's body and the axis of rotation decreases by a factor of two when the skater pulls in her arms,

$$r \rightarrow r' = 0.5 \cdot r$$

then we can compute the increase in angular velocity of the skater:

$$\omega \rightarrow \omega' \sim r'^{-2} = \left(0.5 \cdot r\right)^{-2} = 4r^{-2} \sim 4\omega$$

The new angular velocity will have increased by a factor of four. For example, if she was making two rotations per second with her arms out, she will make eight rotations per second with her arms pulled in.

If you have an office chair that spins well, you can test this for yourself. Start yourself spinning really fast. When you stick your legs out, you will slow down. When you pull them back in, you should speed up again. Play around with this for a while, and when you're done vomiting, you'll have a great appreciation for the intricacies of angular momentum conservation.

The Times, They're a Changin'

Whenever I run at the track, I'm always annoyed that the gym workers make you run in different directions depending on the day of the week. They tell me that changing directions works muscles on both sides of the body evenly, and you don't want to work one side more than another. This may be true, but track athletes always run counterclockwise in meets, and they generally don't have one leg that's cartoonishly more muscular than the other.[80] I wouldn't mind this mild nuisance, but I can never remember which way I'm supposed to be going on any given day. The whole thing never made sense to me until I read about the 2011 earthquake in Japan. The earthquake not only caused extreme devastation but also shortened the length of a day by about one microsecond.[81] Because the quake lifted some parts of the land, the Earth had to slow down in order to conserve angular momentum. In a similar way, running the same way around a track will cause the length of a day to shift because the total angular momentum of you plus the Earth must be conserved. **If everyone in the world ran in the same direction around a track at the North Pole, how much would the length of a day change?**

The curved part of a track is half a circle and takes up about 100 meters in length. Given that the circumference C of a full circle is related to the circle's radius r by the equation

$$C = 2\pi r$$

it must be true that the distance around half a circle is πr, so $r = 32$ meters.

---

80 To be fair, even small imbalances in muscle alignment can lead to injuries.

81 A microsecond is equal to 0.000001 seconds.

There are about $N = 7 \times 10^9$ people in the world. We can compute the angular momentum L of each person using the formula

$$L = mvr$$

where m is the person's mass, v is the velocity, and r is the radius of the circle in which s/he's running. The total angular momentum would then be the average angular momentum multiplied by the world population. I'll assume the values listed in the table below.

NAME	SYMBOL	VALUE
RADIUS	r	32 m
AVERAGE MASS OF A PERSON	m	65 kg (~140 LBS)
RUNNING SPEED	v	8 m/s
NUMBER OF PEOPLE	N	7×10^9 people

Using these values, the total angular momentum is

$$L = Nmvr = \left(7 \times 10^9\right) \cdot (65 \text{ kg}) \cdot (8 \text{ m/s}) \cdot (32 \text{ m}) = 1.2 \times 10^{14} \text{ kg} \cdot \text{m}^2 / \text{s}$$

The angular momentum of the Earth is 7.1×10^{33} kg·m²/s. If everyone gathered at the North Pole and ran around a track, it would change the Earth's angular momentum by 3.5×10^{-18} percent. That's a 0.0000000000000000035 percent change.[82] The length of a day scales with the angular momentum as

$$t_{day} \sim L^{-1}$$

so the day would get 4.4×10^{-18} percent shorter. Although the length of a day would change, the difference would be so small that even the most precise atomic clocks could not measure it. Compared with natural disasters, humans are thoroughly insignificant.

---

82 The details of calculating this are a little complicated, but if you add up the angular momentum values for each atom in the Earth, this is what you'll get.

PHYSICS VI:

Smooth Oscillator

When a weightlifter does a clean and jerk, the ends of the bar bounce up and down. Hit a softball the wrong way and you'll notice the metallic bat vibrating in your hands. Get punched in the face by a boxer and you may feel the skin of your cheeks flap like a flag. Sports are filled with materials that oscillate, vibrate, and jiggle. Objects that exhibit this type of motion have some force that tries to keep them in a stable equilibrium position. If you push the object slightly away from that position, it tries to snap back but often overshoots and then has to snap back the other way. This creates a cyclic pattern of snapping forward and back until friction saps all the energy out. This "snapping" force is due to the elastic properties of the material.

Remarkably, all the different types of jiggling can be approximated very well by a single equation, provided the wobbly motion is not too large. This type of motion is called simple harmonic motion, and the equation is called Hooke's law, after Robert Hooke, who first applied it to elastic springs.[83] The equation can be written as

$$F = -kx$$

where F is the force on the oscillating object, x is the distance the object has been displaced from equilibrium, and k, called the spring constant, is related

---

[83] Strictly speaking (in fact, even if we're not speaking strictly), Hooke's law is not a law. Pretty much every material violates it. I mentioned before how you can approximate any function using something called a Taylor series expansion in polynomials. That's all Hooke's law is. In this case, we're approximating the force caused by a spring as a function of how far it's displaced from equilibrium. We could have written the force function with more terms in the series

$$F(x) = -k_1 x + \frac{1}{2} k_2 x^2 + \frac{1}{6} k_3 x^3 + \ldots$$

where k_1, k_2, k_3, and so forth are all different spring constants. The extra terms are called "nonlinear" terms, but they can all be neglected if the displacement is small enough (e.g., if $x = 0.01$, then x^2 is only 0.0001, x^3 is only 0.000001, etc.).

to the elastic properties of the material. Different objects will have different values of k depending on their geometry and the material they're made out of. The spring constant for a small child on a swing set is about 180 N/m, whereas the spring constant for two atoms bonded together is about 3 N/m.

The time it takes for a springy material to complete one oscillation is called the period T. The frequency f of these oscillations is given by

$$f = \frac{1}{T}$$

and it is generally measured in Hz. For various reasons, physicists prefer to talk about angular frequency:[84]

$$\omega = 2\pi f$$

Most problems involving simple harmonic motion are fairly complicated and require the use of trigonometry. For that reason I've included only one problem on the matter. However, because it's an important topic, I'll describe a few other sport-related examples at the end of the problem.

---

[84] This is partly to touch base with angular velocity. When viewed from the side, an object moving at a constant speed along a circular path looks exactly like one undergoing simple harmonic motion.

Boobs: Nature's Harmonic Oscillator

Tecmo's *Dead or Alive: Xtreme 2* may have fairly mediocre gameplay, but it does have one feature that drew a lot of attention during its release: big bouncing breasts. According to videogame reviewer Douglas Perry, the volleyball game features "independent breast physics which create an asynchronous movement while giving a water balloon sensation to each mammary gland." [85]

I'd like to think the good people of Tecmo devoted an entire lab to testing jiggle physics, but, sadly, the bouncing isn't anywhere close to realistic. Still, there's something about the sway of a woman's breasts that leaves our very souls oscillating, scintillating, and reverberating into a world of happy. **Just how springy is a boob?**

Let's do an experiment. Find the boob closest to you and grab it. [86] Try gently pulling it slightly up and then let go. You'll find it returns to its original position after perhaps undergoing an oscillation or two. Now try gently pulling the boob down. Assuming it's a relatively young boob, it should return to its original position. [87] In this way, a boob is just like a spring. Both have one stable equilibrium position, and if you displace them slightly from this equilibrium, they go boingy, boingy, boingy, boingy.

---

85 Douglas Perry, "Dead or Alive Xtreme 2 Hands-on." IGN. Retrieved July 1, 2011, http://xbox360.ign.com/articles/730/730471p1.html.

86 If the boob in question is not your own, make sure you get permission first. I can't stress this point enough. There's no amount of "It was for science!" that will save you from a swift kick to the groin.

87 Whether you like it or not, boobs really are just like mechanical springs. Over time both lose elasticity and firmness and will not return to their equilibrium position. On a related note, I should mention that all boobs referenced in this problem are over the age of 18.

Left to its own devices, a boob might undergo one period of oscillation in about 0.5 seconds. This would give a frequency

$$f = \frac{1}{T} = 2 \text{ oscillations per second}$$

and an angular frequency,

$$\omega = 2\pi f = 12.6 \text{ radians per second}$$

The "springiness" of a harmonic oscillator is measured quantitatively through the constant k in the Hooke's law equation for the force F as a function of displacement x:

$$F = -kx$$

Larger values of k mean bigger forces (i.e., stiffer springs), and smaller values of k mean weaker forces (i.e., looser springs). The value of k is given by

$$k = m\omega^2$$

where m is the mass of the oscillator. Assuming boobs of mass one kilogram, we find their spring constant to be

$$k = (1.0 \text{ kg}) \cdot (12.6 \text{ Hz})^2 = 160 \text{ kg/s}^2$$

At 160 kg/s², a boob is about as springy as a child on a swing set, though I don't pretend this example provides a whole lot of context. If not for context, then why did we do this problem?[88] It turns out there is a lot of rich sports physics in harmonic oscillators, and I would be remiss if I didn't include at least one problem on this phenomenon. Sadly, the problems tend to be quite involved, so we're only able to scratch the surface here. To compensate for the lack of actual problems, I've included descriptions of three phenomena stemming from harmonic oscillators that are highly relevant to sports: damping, coupling, and resonance.

Damping

Consider, for example, setting our videogame volley boobs in motion. The oscillation will eventually die down. Why? "Air friction!" you say. But even a boob left in a vacuum would not oscillate forever. There are dissipative forces in the boobs themselves that will dampen the motion, but this is not the only method in which energy can be sucked, so to speak, out of the system.

Coupling

We've been neglecting a fairly substantial entity that milks energy out of the oscillation—namely, the other boob. I don't mean to stereotype, but they typically come in pairs, and if we start undulating the first, there's a good chance the second will jiggle shortly thereafter. Because the two are firmly affixed to a torso, some small force can be transmitted through the torso from one boob to the other, causing it to undulate. When this happens, we say the two oscillators are "coupled." Coupled systems transfer energy from one oscillator to

---

88 I mean other than the fact we got to talk about boobs.

another, but I'll leave it to the burlesque dancers to determine exactly how this happens for mammary glands. There is, however, a sports-related example of coupled oscillators: the sweet spot. If you want to hit an upper-deck home run, you better transfer a good portion of your swing energy into the motion of the ball rather than into the vibrations of the bat molecules. Baseball bats and tennis rackets have certain spots where energy is transferred more efficiently to the ball. This happens because of the way the many atoms that make up these objects are coupled to each other. When you hit a sweet spot, the ball leaves with a much higher velocity.

Resonance

Finally, there is the related problem of resonance. If you've ever played on a swing set, you know you can swing quite high if someone pushes you every time you swing back. This happens because the pushing frequency closely matches the natural frequency at which the swing would oscillate by itself. If your friend pushes you forward once per week, you're obviously not going to go very high. Likewise, if your friend pushes you forward ten times per second, you might get displaced, but the amplitude of your swing will still be small. The same is true for boobs or any other harmonic oscillators. Pushed at the right frequency, the oscillation amplitude can grow quite large. This is presumably why you don't see many D-cup marathoners, because the stride frequency closely matches the two-Hz natural frequency of a boob and would likely induce painful large-amplitude bouncing.

PHYSICS VII:
Under Pressure (Buoyancy and Temperature)

We've used density in various calculations throughout the book. Formally, an object's density ρ is obtained by dividing its mass m by the volume it occupies V:

$$\rho = \frac{m}{V}$$

Density is often used to describe what's going on in gases and liquids. For example, the reason a canoe floats is that it's less dense than the water it floats on. Gravity pulls objects to the Earth, but heavier objects are pulled more strongly. If you try to push the canoe under water, some of the denser water will be displaced upward. Gravity fights this, and the end result is that the water will push up on the canoe with a buoyant force, given by

$$F_{buoyant} = \rho V g$$

where ρ is the density of the fluid, V is the volume of fluid that gets displaced, and g is, again, the acceleration of gravity.

Although the canoe example deals specifically with mass, density is a fairly general term describing the amount of something in a given space. The "something" could be anything: mass, electric charge, stars in the sky, population, ears on Roger Clemens's head, etc. Pressure is like a force density. Specifically, it's the amount of force F applied per unit area A:

$$P = \frac{F}{A}$$

We're constantly surrounded by pressure. As we speak, your body has about 50,000 pounds of force exerted on it due to air pressure. "Why aren't I crushed?" you ask incredulously. You aren't crushed for the same reason a basketball stays inflated: there's pressure inside that balances the air on the outside. In the case of the basketball, the pressure inside is slightly larger than the air pressure outside, so the ball stays taut.

The pressure of a gas is well approximated by the ideal gas law,

$$PV = Nk_BT$$

where V is the volume occupied by the gas, N is the number of molecules in the gas, T is the temperature of the gas in units of K, and kB is a constant equal to roughly 1.38×10^{-23} J/K.[89] To pump up a basketball, you push extra air molecules inside. Because the temperature doesn't change and the volume stays almost the same, the pressure must increase. In contrast, people who race hot air balloons keep the pressure and volume of the balloon constant but increase the temperature so the number of molecules inside decreases until the balloon is lighter than air. Whenever you change one variable in the equation, one or more other variables will adjust to compensate so as to ensure that the ideal gas relationship holds.

[89] The "K" in the temperature unit stands for "Kelvin" and is named after William Thomson, First Baron Kelvin. You can convert between Kelvin and Celsius via the equation

$$[^{\circ}C] = [K] - 273.15.$$

Roughly speaking, the temperature of a substance corresponds to how fast the atoms inside are moving. Whenever you change the temperature of a substance by an amount T, you're altering the energy stored inside by an amount

$$\Delta E = C_P m \Delta T$$

where m is the mass of the substance, and C_P, called the heat capacity, measures an object's capacity for storing thermal energy.[90] The bigger the heat capacity, the more energy you need in order to raise an object's temperature.

In this section are four examples illustrating the effects of pressure, buoyancy, and temperature.

---

[90] The subscript "P" in the head capacity indicates the value is taken at constant pressure.

Breaking the Sound Barrier for Under $200

In a *South Park* episode titled "The China Problem," Cartman is troubled after watching the opening ceremonies of the 2008 Beijing Olympics and believes China will invade the United States at any moment. If the opening ceremonies were scary, then China's performance in table tennis was downright mortifying.[91] China took gold, silver, and bronze in men's singles, then took gold, silver, and bronze again in women's singles before taking gold in both men's and women's team table tennis. Is there any way the United States can stop the juggernaut that is the Chinese Ping-Pong team?

Although having massively juiced-up ping-pong players might work *and* be hilarious, there is another solution: air pressure. **How fast can you shoot a Ping-Pong ball using air pressure alone?**

If you've been following the book so far, you may be disappointed that I've yet to provide you with any practical applications. Fear not, for this problem will teach you how to break the sound barrier for under $200! Just follow these simple steps and you're on your way to a supersonic Ping-Pong battle:

Step 1. It's time to go shopping. You'll want a two-inch wide, three-meter long piece of PVC pipe, Teflon tape, a vacuum pump adapter, and some packing tape from your local hardware store. This should run you at most $30 to $40. You'll also need a Ping-Pong ball or, better yet, several. A package of 100 should only cost around $10. Finally—and this is the most expensive part—you'll need a good vacuum pump. You can find one used for cheap, but even a halfway decent new one should only cost around $150. The total cost of all materials should be under $200.

---

91 Ping-Pong hasn't always been a harbinger of doom for America. In 1971, a chance meeting between American and Chinese Ping-Pong players Glenn Cowan and Zhuang Zedong paved the way for Nixon's visit to China, which opened Chinese-US relations. At its core, sports are really just a bunch of humans pushing a ball around in various ways, but this is a prime example of how, at times, it can mean so much more.

Step 2. Drill a hole about one foot up the PVC pipe and insert the vacuum pump adapter. Make sure it's a tight fit so no air can escape. You may need to wrap some Teflon tape around the adapter to ensure a tight seal. Once you've inserted the adapter, roll the Ping-Pong ball down the long end of the PVC pipe. The adapter should be long enough to prevent the ball from falling all the way through.

Step 3. Seal off both ends of the PVC pipe with the packing tape. Make sure no air can leak into the pipe. You may need to sand the ends down to ensure a good seal.

Step 4. Hook the vacuum pump to the adapter and evacuate the air from the PVC pipe. If you've set it up correctly, you should see the packing tape start to bow inward once the vacuum pump is switched on.

Step 5. Poke a large hole in the packing tape on the side closest to the adapter. This will reintroduce the air from one side, thus pushing the Ping-Pong ball out through the packing tape on the other side. **How fast will the ball be traveling when it exits the pipe?**

When air is reintroduced to the pipe, the Ping-Pong ball feels the force of air pressure only on one side. This force can be written as

$$F = PA$$

where $P = 1.0$ atm is the pressure of the atmosphere under normal conditions, and $A = 13$ square centimeters is the cross-sectional area of the Ping-Pong ball.

This force will act on the Ping-Pong ball for a distance $x = 2.7$ meters, which is equal to the distance that the ball travels in the pipe. The energy absorbed by the Ping-Pong ball is given by

$$E = Fx = PAx$$

where we've plugged in the expression for the force provided above.

The work done on the Ping-Pong ball must be equal to the rise in kinetic energy:

$$KE = \frac{1}{2}mv^2$$

By setting these two expressions equal, we can solve for the velocity of the Ping-Pong ball as it leaves the pipe:

$$v = \sqrt{\frac{2PAx}{m}}$$

I'll assume the following values:

NAME	SYMBOL	VALUE
PRESSURE	P	1.0 atm
AREA OF THE PING-PONG BALL	A	13 cm²
DISTANCE TRAVELED	x	2.7 m
MASS OF THE PING-PONG BALL	m	2.7 G

Plugging in, we get

$$v = \sqrt{\frac{2 \cdot (1.0 \text{ atm}) \cdot (13 \text{ cm}^2) \cdot (2.7 \text{ m})}{(2.7 \text{ g})}} = 510 \text{ m/s}$$

That's 510 m/s or roughly 1.5 times the speed of sound![92] With speeds like this, the Chinese Ping-Pong team won't stand a chance.

---

[92] Because of air friction, the speed of the Ping-Pong ball dies off pretty quickly once it leaves the pipe. Still, it's coming out fast enough to be dangerous, so don't stand in front of it.

Slam-Your-Head-on-the-Radiator Nerf Basketball

To all the young athletes out there, whenever you're doing plyometrics or any other jumping exercise, it's a good idea to land softly. This will save your joints from some very painful problems later on. The Muscleless Wonder has some great advice for this: "Imagine you're playing Slam-Your-Head-on-the-Radiator Nerf Basketball." At some point, every kid's played Slam-Your-Head-on-the-Radiator Nerf Basketball. You're upstairs in your room, trying to "be like Mike" and going up for a dunk on the seven-foot high rim attached to your bedroom door. Inevitably, you hear your mom yell, "Stop horsing around up there before you crack your head on the radiator!" Do you stop playing? Of course not! You just try to land very lightly so Mom can't hear you.

Landing softly is about absorbing energy.[93] The kinetic energy of the body gets diverted into many other forms: elastic energy in your muscles, vibrational energy in the floorboards, acoustic energy in your mom's ears, etc. Eventually, all energy goes to heat. If you convert it to heat directly, rather than letting it pass through your joints and vital organs first, then in principle you will be safe jumping from any height.[94] **If you dissipated all the energy after jumping off the Empire State Building, how much would your temperature rise?**

---

93 If you've ever done an egg-drop contest in a high school physics class, you know what I'm talking about. In this contest, students must drop an egg from a window on the top floor without it breaking. I once had a student who placed the egg inside a whole chicken and argued, "It's nature's way of protecting eggs." He really should have defrosted the chicken.

94 This is very much a practical problem for the elderly. With brittle bones, it takes much less energy for a hip to snap. You can actually find padded pants online designed to create just this energy-dissipating effect.

The total energy initially comes from gravitational potential energy,

$$GPE = mgh$$

The Empire State Building is 380 meters (~1,250 feet) high. Assuming an 80-kilogram (~180-pound) athlete, the total energy to be dissipated in the landing is

$$E = mgh = (80 \text{ kg}) \cdot (9.8 \text{ m/s}^2) \cdot (380 \text{ m}) = 3.0 \times 10^5 \text{ J}$$

That's 300,000 J of energy you need to dissipate efficiently to stay alive.

The heat capacity of water is $C^P = 4180$ J/kg·°C. Because our bodies are composed mostly of water, they likely have a similar heat capacity. Assuming this to be the case, we can calculate the total change in temperature induced by an extra 300,000 J of energy by solving the heat capacity equation for temperature:

$$\Delta T = \frac{E}{mC_P} = \frac{(3.0 \times 10^5 \text{ J})}{(80 \text{ kg}) \cdot (4180 \text{ J/kg} \cdot °C)} = 1.0 \text{ °C}$$

Even if all the energy dissipated directly into heat, jumping off the Empire State Building would only increase your body temperature by 1.0°C. Unfortunately, the energy will not go directly to heat, as some of it will take a detour in the form of broken bones and ruptured spleens. If you think you've got the hang of this one, try calculating how much your body temperature could rise if you jumped off Mount Everest.[95]

---

95 By my estimate, this should be about 20˚C. Note that both this and the estimate for the Empire State Building assume that all the energy goes into heating your body as opposed to the air, ground, and other objects.

How to Chuck a Prolate Spheroid

There are many impressive athletic feats posted on YouTube. Jarron Gilbert jumping out of a pool, Evan Longoria snatching a foul ball out of the air with his bare hand, and the Muscleless Wonder dropping a 220-pound barbell from overhead onto 400 condiment packets and getting splattered by the resulting explosion are just a few that come to mind. One video in particular caught my eye: Michael Vick throwing a football out of a stadium. No one will refute that NFL quarterbacks have great arms, but is there any way for us normal Joes to increase our range? **How much farther can you throw a football filled with helium?** [96]

---

96 For simplicity, I'm ignoring the force of air friction. As you'll see in a moment, including air friction would change the result.

Before we start, I should mention that *MythBusters* tested this myth in an episode titled "Catching a Bullet with Your Teeth and Helium Footballs." Their results showed that there's not much of a difference between helium and air-filled balls, and, if anything, air-filled balls travel further. I'll discuss why in a moment, but for now, let's just consider how the added buoyancy of the ball affects the motion.

The range equation says that the maximum range R of a projectile can be computed using the equation:

$$R = v^2 / a$$

where v is the velocity and a is the acceleration downward. Most of the time, we set a equal to the acceleration of gravity $g = 9.8$ m/s^2, but rigorously speaking, this is only true in vacuum. In truth, there are at least two forces that contribute to the acceleration of the ball: gravity and buoyancy. For a normal football, the gravitational force must take into account the mass of both the leather and the air inside the ball. This force can be written as

$$F_{gravity} = -\left(m_{skin} + m_{air}\right)g$$

where m_{skin} is the mass of the "skin" of the ball and m_{air} is the mass of the air inside the ball, which can be expressed as

$$m_{air} = \rho_{air}V$$

Here, V is the volume of the ball.[96] The minus sign in the force is just there to remind us that the force is downward. In addition to gravity, we also have to consider the buoyant force of the air pushing up on the football:

$$F_{buoyant} = \rho_{air}Vg$$

---

96 I'm assuming the volume of the skin is much smaller than the volume occupied by the air inside.

We can compute the total force on the ball by adding the downward gravitational force to the upward buoyant force. In doing so, the buoyant force will cancel the gravitational force on the air inside the ball, leaving

$$F_{total} = -m_{skin}g$$

A similar analysis can be presented for the helium-filled ball. In this case, the mass of helium will replace the mass of air in the gravitational force equation. The total force is still the sum of the two, but now buoyant force will not cancel out:

$$F_{total\ helium} = F_{buoyant} - F_{gravity}$$
$$= \rho_{air}Vg - \rho_{helium}Vg - m_{skin}g$$

where ρ_{helium} is the density of helium.

For a helium balloon, the buoyant force is greater than the gravitational force on the balloon skin, so the balloon rises up. For a football, the skin will be massive enough that the helium-filled ball will still fall to the Earth, but it will do so with less acceleration than the air-filled balloon. Because acceleration is proportional to force, the ratio of the accelerations for the helium-and air-filled balloons is equal to the ratio of their forces:

$$\frac{a_{helium}}{a_{air}} = \frac{F_{total\ helium}}{F_{total\ air}} = \frac{\left(\rho_{air} - \rho_{helium}\right)V - m_{skin}}{m_{skin}}$$

Using the assumed values listed in the table, we find

$$\frac{a_{helium}}{a_{air}} \approx 0.987$$

or, put more simply, a helium-filled balloon will accelerate downward at only 98.7 percent the acceleration of an air-filled ball.

From the range equation, we see that

$$R \sim a^{-1}$$

Using scaling, we can estimate that with a helium-filled football, you could throw

$$R \rightarrow R' \sim a'^{-1} = \left(0.97 \cdot a\right)^{-1} = 1.03 \cdot a^{-1} \sim 1.013 \cdot R$$

or roughly 1.3 percent farther. To put this figure in perspective, let's consider what it would mean for an NFL quarterback. Under ideal conditions, top NFL quarterbacks can throw a football about 80 yards. A 1.3 percent increase would bump this up to 81 yards. This is significant, but it's not likely to take you from D3 backup to NFL prospect. Moreover, we still have the problem that *MythBusters* actually tested this and got the opposite result. What gives? The problem is that our calculation ignored the role of frictional drag. Without air friction, there's no force to slow the forward motion of the ball, and buoyancy becomes relatively important. However, in the real world, drag exists and is a larger force than buoyancy. Because a force can decelerate a smaller mass more easily, the helium-filled ball will travel a shorter distance than the more massive, air-filled ball. We'll discuss how to calculate the drag force in a moment, but for now let's treat this as a cautionary tale. Although we can model the world as being frictionless, we should always go back and check to see if adding friction significantly changes our result. Also, it's generally a bad idea to disagree with guys who blow things up for a living.

NAME	SYMBOL	VALUE
DENSITY OF AIR	ρ_{air}	1.2 kg/m³
DENSITY OF HELIUM	ρ_{helium}	0.17 kg/m³
VOLUME OF A FOOTBALL[97]	V 5000	cm³
MASS OF A FOOTBALL	M_{skin}	0.4 kg

[97] In geometric terms a football is a prolate spheroid with volume a $V = 4\pi a^2 b,/3$ where a is the equatorial radius and b is the polar radius. To calculate the volume, I chose $a = 3.5$ inches and $b = 6$ in.

Jordan's on Fire

When an NBA player hits a few shots in a row, the announcers often say he's on fire. The makers of the 1993 video game *NBA Jam* took this expression a little too seriously by creating players who could, quite literally, catch on fire. While it's nice to be on a hot streak, morphing into the Human Torch is not necessarily beneficial to a basketball player. **How much would the ball expand if it were on fire?**

A regulation basketball is inflated with a pressure of about eight pounds per square inch, which is about half of atmospheric pressure. As you increase the temperature of the gas inside the ball, the pressure will increase as well, but if you make the ball out of a very stretchy material, the volume may increase instead. To simplify the problem, I'm going to assume the ball is made of a rubber that stretches easily so that the pressure inside will remain roughly constant. I'm further going to assume the rubber doesn't ignite.

According to the ideal gas law,

$$PV = Nk_B T$$

from which we can see the temperature scales with the radius of the ball as

$$T \sim V \sim R^3$$

or, equivalently,

$$R \sim T^{1/3}$$

Room temperature is about 300 K (~25°C). The phrase "on fire" is a little undefined, as fire comes in a wide variety of forms, but I'll assume a temperature of 2800 K (~2500°C), which is roughly the temperature of a blowtorch flame. At 2800 K, the temperature has increased by a factor of about 9.3:

$$T \rightarrow T' = 9.3\ T$$

which means that the radius will increase as

$$R \rightarrow R' \sim T'^{1/3} = \left(9.3\ T\right)^{1/3} = 2.1\ R.$$

A basketball, which is normally 12 centimeters in radius, would now be 2.1 times larger, or roughly 25.2 centimeters in radius. This is even bigger than the 22.9-centimeter rim radius. Clearly, being on fire would not cause a hot streak.

PHYSICS VIII:
Only Heiden Can Skate on a Frictionless Surface

In the physics world, we often assume friction is absent or, at the very least, negligible. We do this for several reasons. First, it's often a good assumption to make given the time scales over which you're observing a system in a physics lab. Second, it allows you to contemplate the fundamental physical forces at play rather than complexities that obscure the important results. The third reason is, without question, the most important of all: it makes solving problems easier.

Unfortunately, the sports world has very little to do with fundamental physics, and friction often rears its ugly head as a necessary part of the problem. In cases like this, we can approximate the frictional force on an object, also known as the drag force, using the equation[98]

$$F = 6\pi\eta r v + \frac{1}{2}C_D\rho A v^2$$

where η (a Greek letter pronounced "eta") is called viscosity and measures how poorly a fluid flows, r is the object's radius, C_D is a constant called the drag coefficient, ρ is the density of the fluid, A is the cross sectional of the object, and v is the velocity at which it's traveling.[99] In principle, one should

---

[98] There's a scene in the movie *National Lampoon's European Vacation* in which Clark Griswold and family get stuck in a London roundabout and end up circling around for hours. At every pass, Chevy Chase as Clark Griswold exclaims, "Hey, look, kids! There's Big Ben! There's Parliament!" I feel a little like Griswold here, because the drag force is yet another example of a Taylor series. Once again, there's no fundamental physics here per se. We've just approximated the force as a function of the velocity by writing it as the sum of polynomials. I should note that the first term in teh drag force is known as Stokes' Law and is true only for spheres. Objects of different shapes will have a different factor in front, but will retain the same linear dependence on velocity.

[99] I'm approximating our object as spherical, which is usually a decent assumption if all you want is an order of magnitude estimate.

really include both of these terms in calculations of the drag force, but in practice it's often sufficient to omit one of them. When the velocity v is small, we can neglect the second term; when it's large, we can neglect the first. You can see this by looking at how the terms scale. If you increase the velocity by a factor of ten, the first term increases by 10, but the second increases by 100. Similarly, decreasing the velocity by a factor of ten will decrease the first and second terms by factors of 10 and 100, respectively. The ratio of the two terms determines the cutoff for what constitutes large or small.[100]

In addition to creating drag on a moving object, air friction is also responsible for curving the flight of spinning baseballs. The "curving" force arises because of the way a spinning ball throws air off itself. As the ball passes through the air, some of the air molecules get pulled with the rotation of the ball, whereas others get deflected. The net effect is that air molecules on one side of a spinning ball are deflected more than those on the other side, and by Newton's third law, there is a reaction force on the ball. This phenomenon is called the Magnus Effect, after physicist Heinrich Magnus, who discovered it in 1852. If the axis of rotation is perpendicular to the velocity, then the force on the ball can be calculated using the equation

$$F = S\omega v$$

where S is a quantity related to the air resistance across the ball's surface, ω is the ball's angular velocity, and v is the ball's speed.

Although friction is certainly a pest for physicists, perhaps it's not so bad for the sports world. After all, if friction didn't exist, baseball stadiums would double in size to keep the number of home runs down, football concussions would skyrocket because there'd be almost no limit to the speed at which players could sprint, and curveballs would be a thing of the past. The following examples illustrate the role friction plays in sports.

---

100 This is known as the Reynolds number Re, which is defined as

$$R_e = \frac{\rho v r}{\eta}$$

For $Re < 1$, the first term is sufficient. For $Re > 1,000$, the second term is sufficient.

Check Out the Cannon on That Fireballer

I still remember the T-shirts sold outside Fenway Park when Johnny Damon left to sign with the Yankees:

Looks like Jesus

Acts like Judas

Throws like Mary.

Sexism aside, the sentiment was quite true. Anyone who watched Damon knew he needed a cutoff man to hit the cutoff man. Although Damon is a poor example, there are a handful of outfielders who can reach home plate on the fly. These men are typically said to "possess a cannon" and are often called "fireballers." **What's a bigger exaggeration—having a cannon or being a fireballer?**

The Physics Fact Book lists artillery shell speeds in the range of 440 to 1,667 m/s. We can assume cannonball speeds lie comfortably within this range.

A baseball will catch on fire if you hold it in a candle flame long enough. Candles generally burn at about 1,000°C. From this, we can conclude that to be taken literally, a fireballer's throw must be fast enough to heat the ball to around this temperature. If you've ever observed a meteorite burning up as it descends into the Earth's atmosphere, then you know this is, at least in principle, possible.

Besides gravity, which will be a small effect here, the force on a fast-moving baseball is caused by air friction. This drag force is given by

$$F = \frac{1}{2}C_D\rho Av^2$$

where C_D is the drag coefficient, ρ is the density of the air, A is the cross-sectional area of the ball, and v is the velocity of the ball. Using the definitions of power and velocity, we can show that the rate at which energy is dissipated by the ball due to air friction is given by

$$P = \frac{\Delta E}{\Delta t} = Fv = \frac{1}{2} C_D \rho A v^3$$

where ΔE is the energy lost in a time interval Δt. The energy lost will inevitably end up as heat, some of which will raise the temperature of the ball. Assuming all of the energy goes into heating the baseball, the ball's temperature increase can be determined by the specific heat equation,

$$\Delta E = C_p m \Delta T$$

where C_p is the specific heat and m is the mass. From this and the power equation, we can solve for the velocity needed to induce combustion:

$$v = \sqrt[3]{\frac{2 C_p m \Delta T}{C_D \rho A \Delta t}}$$

Using the assumed values listed in the table, we find the velocity needed to ignite the baseball is

$$v = \sqrt[3]{\frac{2 \cdot (3470 \text{ J/kg} \cdot {}^\circ\text{C}) \cdot (145 \text{ g}) \cdot (975 \text{ }^\circ\text{C})}{(0.4) \cdot (1.2 \text{ kg/m}^3) \cdot (44 \text{ cm}^2) \cdot (1.0 \text{ s})}} = 870 \text{ m/s}$$

or roughly 2.5 times the speed of sound. In conclusion, whether or not it's better to be a fireballer or have a cannon very much depends on the quality of the cannon.

NAME	SYMBOL	VALUE
AREA	A	44 cm²
DRAG COEFFICIENT	C_D	0.4
HEAT CAPACITY	C_P	3470 J/kg °C
MASS	m	145 g
AIR DENSITY	ρ	1.2 kg/m3
TEMPERATURE CHANGE	ΔT	975°C
TIME	Δt	1.1 s

Baseball's a Drag

Whenever I hear of some top pitching prospect lighting up the radar gun, I always wonder the same thing: "Where was that measurement taken?" Because of air friction, the speed of a pitch drops during the 0.4 or so seconds the ball is in flight. A measurement taken right when the ball leaves the pitcher's hand will give a faster reading than one taken when the same ball crosses the plate. Because of friction, the speed must drop, but by how much? What does a radar gun really measure? **How much the does the speed of a fastball actually drop?**

Aroldis Chapman's record-breaking fastball was measured at 105 mph, so we can use this as an upper bound. No matter where this speed was measured, we know that the speed at release must be at least this fast because that is the point at which the ball is traveling fastest. For the time being, I'll assume the speed at release is 105 mph.

The force friction on a baseball is given by the equation

$$F = \frac{1}{2} C_D \rho A v^2$$

where C_D is the drag coefficient, ρ is the density of the air, A is the cross-sectional area of the ball, and v is the velocity of the ball. Typical values for these parameters are listed in the table below:

NAME	SYMBOL	VALUE
DRAG COEFFICIENT	C_D	0.4 for a sphere
AIR DENSITY	ρ	1.2–1.3 kg/m^3
CROSS-SECTIONAL AREA	A	4.2×10^{-3} m^2
MASS OF A BASEBALL	m	0.145 kg

Knowing the mass m of a baseball, you might think we could just solve for the acceleration using Newton's second law and then use this acceleration to find the change in velocity. The problem is that the velocity in the force equation is changing, so the force is also changing. The question is still solvable, but you need to use the equation

$$v_f = \frac{v_i}{1 + v_i C_D \rho A t / 2m}$$

where v_i and v_f are the initial and final velocities, respectively.[101] Assuming the pitch is in flight for 0.45 second, the final velocity is

$$v_f = \frac{(105 \text{ mph})}{1 + (105 \text{ mph}) \cdot (0.4) \cdot (1.25 \text{ kg/m}^3) \cdot (4.2 \times 10^{-3} \text{ m}^2) \cdot (0.4 \text{ s}) / \left[2 \cdot (0.145 \text{ kg}) \right]} = 92 \text{ mph}$$

A pitch starting at 105 mph arrives at the plate traveling only 92 mph. If you assume a more reasonable 95-mph fastball, you're down to 84.5 mph by the time it reaches the plate. In essence, you go from being Tim Lincecum to Tim Wakefield in 0.4 seconds. Clearly the difference between being labeled a fireballer and a knuckleballer can have a lot to do with exactly how the radar gun is used. The results are even more startling if you assume Aroldis Chapman's pitch arrived at the plate traveling 105 mph. This means that it would need to have left his hand traveling 140 mph!

---

101 I won't bore you with the details of where this equation comes from, but it's easy to derive with a little bit of calculus knowledge.

Michael Phelpsie Sleeps with the Fishes

Michael Phelps's pursuit of eight gold medals in the 2008 Summer Olympics captivated Americans. The real-life Aquaman didn't disappoint, winning every event in which he competed. It's too bad the *Sopranos* ended the year before, because Phelps could have made a guest appearance. He also could have answered this intriguing riddle: **Could Michael Phelps swim with a pair of cement loafers?**

In the Beijing Olympics, Phelps's 200-meter freestyle time was 102.96 seconds, meaning he was traveling about 1.9 m/s. The reason Phelps can't swim any faster is that the water creates a large frictional drag force that holds him back. We can calculate the magnitude of this force using the drag equation

$$F = \frac{1}{2} C_D \rho_{water} A v^2$$

where C_D is the drag coefficient for a swimmer, ρ_{water} is the density of water, A is the cross-sectional area of the swimmer, and v is the swimmer's velocity. Because Phelps moves at a roughly constant velocity, we know there can't be any net force on him. As such, the force with which Phelps propels himself through the water must be equal to and in the opposite direction of the drag force.

Trying to stay afloat with cement loafers is very different from doing the breaststroke in a swim meet, but for simplicity I'm going to assume Phelps's swimming force is the same in both cases. To stay alive, Phelps's swimming would have to be greater than or at least equal to the downward force caused by the cement shoes. In principle, gravity is pulling down on the cement with a force $F = mg$, but the net force down will be less because of the water's buoyancy. The actual force down will be given by the equation

$$F = \left(\rho_{concrete} - \rho_{water} \right) Vg$$

where $\rho_{concrete}$ is the density of concrete, V is the total volume of concrete, and g is the gravitational constant.

I'll assume the values listed in the table below.

NAME	SYMBOL	VALUE
DRAG COEFFICIENT	C_D	~0.7 for a swimmer[102]
WATER DENSITY	ρ_{water}	1.0 g/cm³
CROSS-SECTIONAL AREA	A	~0.1 m² for a swimmer
CONCRETE DENSITY	$\rho_{concrete}$	2.3 g/cm³
VOLUME OF CONCRETE	V	2.0 ft³
GRAVITATIONAL ACCELERATION	g	-9.8 m/s²

[102] Typical values for the drag coefficient C_D are within an order of magnitude of one, but I cheated a little bit on this one. I've pulled this value from an article by A. Belleman titled "Drag Force Exerted on the Human Body," published in the *American Journal of Physics* 49 (1981): 367–368.

From these values, we can compute both Phelps's swimming force,

$$F = \frac{1}{2}C_D \rho_{water} A v^2 = \frac{1}{2}(0.7) \cdot (1.0 \text{ g/cm}^3) \cdot (0.1 \text{ m}^2) \cdot (1.9 \text{ m/s})^2 = 130 \text{ N}$$

and the force downward from the added weight,[103]

$$F_t = (\rho_{concrete} - \rho_{water})Vg = [(2.3 \text{ g/cm}^3) - (1.0 \text{ g/cm}^3)] \cdot (2 \text{ ft}^3) \cdot (9.8 \text{ m/s}^2) = 720 \text{ N}$$

Thus, Michael Phelps would need to swim about six times harder to survive with cement loafers.

---

[103] Because the human body has roughly the density of water, the buoyant force will nearly cancel out the force of gravity of Phelps's body, so I'll neglect this force here.

House of the Rising Fastball

Stand behind a tree 60 feet away, and I'll whomp you with an optical illusion!

—**Pitcher Dizzy Dean**

It's strange that some people still think a curveball is just an optical illusion. Any golfer with a bad slice will tell you that curves are very much real. One variant on the curveball that has received some attention recently is the so-called "rising fastball." **Can a fastball ever rise up?**

For a fastball to rise, there must be an upward force that's at least as large as the downward gravitational force given by

$$F = mg$$

This upward force is the same force the makes a curveball curve. A typical major league curveball may drop a few inches more than gravity alone would dictate. To set some bounds, most people would probably agree that the actual drop lies between 0.3 inches and 30 inches, so 3 inches seems reasonable. From this number, we can use Newton's second law,

$$a = \frac{F}{m}$$

and one of the equations of motion,

$$h = \frac{at^2}{2} = \frac{Ft^2}{2m}$$

to calculate the force responsible for the curve:

$$F = \frac{2mh}{t^2}$$

This force will be equal to the Magnus force caused by the rotation of the ball,

$$F = S\omega v$$

where S is constant related to the air resistance across the ball's surface, ω is the angular velocity, and v is the speed of the ball. The angular velocity is 2p times the number of rotations the ball makes per second. Judging from slow-motion videos of curveballs, they appear to rotate about ten times during the 0.5-second pitch, giving an angular velocity of about 130 Hz. Setting the Magnus force equal to the force responsible for the curveball's drop, we can solve for the value of S:

$$S = \frac{2mh}{\omega v t^2} = \frac{2 \cdot (0.145 \text{ kg}) \cdot (3 \text{ inches})}{(130 \text{ Hz}) \cdot (80 \text{ m/s}) \cdot (0.5 \text{ s})^2} = 19 \text{ mg}$$

With this value in hand, we can estimate the minimum angular velocity needed to make a fastball rise by setting the gravitational force equal to the Magnus force and solving for the angular velocity:

$$\omega = \frac{mg}{Sv} = \frac{(0.145 \text{ kg}) \cdot (9.8 \text{ m/s}^2)}{(19 \text{ mg}) \cdot (80 \text{ mph})} = 2000 \text{ Hz}$$

With an angular velocity of 2,000 Hz (~300 rpm), a rising fastball would require a backspin about 15 times greater than the spin on a curveball.

NAME	SYMBOL	VALUE
MASS OF A BASEBALL	C_D	0.145 kg
GRAVITATIONAL ACCELERATION	g	9.8 m/s²
DISTANCE DROPPED BY A CURVEBALL	h	3.0 in
ANGULAR VELOCITY OF A CURVEBALL	ω	130 Hz
VELOCITY OF A CURVEBALL	v	80 mph
FLIGHT TIME OF A CURVEBALL	t	0.5 s

Sumo Dimensions

One New Years Eve, my wife's family and I watched sumo wrestling on TV while waiting for the ball to drop.[104] There's something quite hilarious about watching large men in diapers run full speed at each other. Perhaps not surprisingly, rikishi have not typically fared well in mixed martial arts. Short of smothering an opponent with their bellies, there's not much sumo wrestlers can do to make rivals submit, but perhaps the big belly gives sumo wrestlers an edge in a street fight. **How thick a layer of fat would a sumo wrestler need to stop a bullet?**

---

104 No, I do not get out much.

There are two ways to attempt this problem: the quick and dirty way and the precise way.

Quick and Dirty Way

Assuming body fat can be considered a fairly viscous fluid, you might correctly suspect that the distance the bullet travels depends on the viscosity η. You may further surmise that the distance the bullet travels will be greater for faster and denser bullets, which suggests the answer will depend on the bullet's initial speed v_i, mass m, and characteristic length r. We can use dimensional analysis to try to guess the answer. The simplest way we can combine all these variables to give an answer with units of distance is

$$\Delta x \sim \frac{mv_i}{\eta r}$$

Plugging in numbers from the table below, we find the bullet will travel about 2.9 meters. This seems a little large. After all, bullets routinely get stuck inside human bodies, so one would expect the actual result to be much smaller than this. As you'll see in a moment, the functional form is correct, but we're off by a dimensionless factor. To find that factor, we'll need to be more precise.

The More Precise Way

We can use the drag force to compute the force on a bullet traveling through fat in almost the same way we did with the "Baseball's a Drag" problem. The only difference is that fat is much thicker than air, so viscous effects will play a large role. As such, it's better to use the part of the drag force that scales as v rather than the part the scales as v^2. The new drag force is written as[105]

$$F = 6\pi\eta r v$$

We can solve for the final velocity as a function of time t by using Newton's second law as we did in the "House of the Rising Fastball" problem:[106]

$$v(t) = v_i \exp\left(-\frac{6\pi\eta r}{m}t\right)$$

In this example, the final velocity should be zero because the bullet has stopped, but this equation predicts the velocity should asymptotically approach zero over time but never actually reach it. What's going on? The bullet loses speed very quickly. So quickly, in fact, that the distance the bullet traveled will still only produce a finite value even though our equation predicts the bullet will be moving from now to eternity.[107] We can use the definition of $v(t)$ to find the distance traveled:[108]

$$\Delta x = \frac{mv_i}{6\pi\eta r}$$

Using the assumed values from the table below, we can estimate:

$$\Delta x = \frac{(15\text{ g})\cdot(965\text{ m/s})}{6\pi\cdot(1000\text{ Pa}\cdot\text{s})\cdot(0.5\text{ cm})} = 15\text{ cm}$$

---

105 To be precise, the bullet is moving at intermediate speeds (Reynold's number ~ 4) in which both viscous and inertial effects are important, so we should incorporate both the v and v^2 terms. Unfortunately, this makes the problem even more difficult to solve. To keep things relatively easy, I'm keeping only the linear term, which produces correct results to within an order of magnitude.

106 As with "Baseball's a Drag," we have to use calculus to solve this.

107 If this seems like a contradiction, imagine it this way. Each second, the bullet moves half the distance to the finish line. As time passes, it keeps getting closer, but it will never actually reach its final destination. For more information on this, look up Zeno's paradox of Achilles and the tortoise.

108 Once again, calculus has come to save the day.

Thus, a sumo wrestler would need 15 centimeters (~0.5 feet) of fat covering his vital organs in order to stop a bullet.

You may have wondered why we took a brief excursion into the realm of dimensional analysis, since the result we got seemed unreasonable. There's a lesson to be learned here. Although it didn't give us a precise numerical answer, it still picked out the correct functional form. As I wrote earlier, physicists tend to be more interested in how a physical quantity depends on other variables than they are in calculating any specific numbers. To understand the physics correctly, it's more important to know how the distance the bullet travels changes with each of the variables than it is to know there's an extra factor of 6π in the equation. If all you want to do is understand the physics, then dimensional analysis gives you what you need for a lot less work. If, for some reason, you need to know more than just the physics, dimensional analysis still provides a nice way of checking the more precise result.

NAME	SYMBOL	VALUE
BULLET MASS	m	15 g
BULLET VELOCITY	a	965 m/s
BULLET RADIUS	r	0.5 cm
VISCOSITY OF FAT[109]	η	1000 pa·s

---

[109] The number cited here is actually the viscosity of lard, which is the most similar substance to human fat for which I was able to find a value of its viscosity in under five minutes.

THAT'S HOW THE BALL BOUNCES

You don't need to be a Vegas bookie to know that sports are largely games of chance. If you're going to play the odds, it's good to know a thing or two about probability. From free throw percentages to batting averages, probability is the language of sports, which is perhaps why gambling on game outcomes is so prevalent. Even if you're not betting money, there's a decent chance you belong to a fantasy league. If so, you're almost certainly drafting the players you believe are most likely to score points over the season. Whether you're a gambling guru, a fantasy fanatic, or just a casual observer, if you follow sports, you're going hear about probability.

At the most basic level, probability is about predicting what you expect to happen given a variety of unknown variables. If you paid any attention in math class, you know the probability of getting heads on a coin toss is 1/2 = 50 percent, and the probability of rolling a six on one die is 1/6 = 16.7 percent. Only slightly harder to find is the probability of tossing heads and rolling a six at the same time. There are 12 equally likely possible outcomes (heads-1, tails-1, heads-2, tails-2, etc.), meaning there's a 1/12, or 8.3 percent, probability of arriving at this combination. To find the probability of two independent events occurring, you simply multiply the probabilities of each event together. For the previous example, we'd combine the 1/2 probability of the coin flip and the 1/6 probability of the die roll to get

$$P = \left(\frac{1}{2}\right) \cdot \left(\frac{1}{6}\right) = \frac{1}{12}$$

Of course, things can get a lot more complicated than this. It helps to know about factorials and the binomial distribution, but I'll talk more about those in a moment. In this section are several problems that illustrate how to think about probability and chance.

Battle of the Time Zones

Fans of National League Central teams have the right to be angry. In 1998, Bud Selig moved the Milwaukee Brewers from the American to the National League, generating an imbalance in the number of teams from each league. Moreover, the Brewers were added to the NL Central Division, making it the only division in baseball with six teams. In contrast, the AL West contains only four teams, giving these teams a significantly higher probability of making the playoffs. **How much of an advantage do AL West teams have over NL Central teams?**

Under the current format, 8 out of 30 teams make the playoffs, which should give each team a playoff probability of about 27 percent. Over a ten-year period, teams should average about 2.7 playoff appearances.

There are 14 teams in the American League and only 4 teams in the AL West, giving these teams a ¼, or 25 percent, chance of winning the division and a 1/11, or 9.1 percent, chance of winning wild card. This means an AL West team has a total playoff probability of about 32 percent. That's 5 percent better than average. Over a ten-year span, AL West teams should make the playoffs about 3.2 times.

In the National League, there are 16 teams, with 6 of these teams in the Central Division. This gives NL Central teams a 1/6 = 16.6 percent, probability of winning the division and a 7.6 percent chance of winning the wild card, for a total playoff probability of about 24 percent. Over a ten-year span, NL Central teams will be playoff bound only 2.4 times on average. That's 3 percent off the mean and 8 percent lower than AL West teams. You might think that 8 percent doesn't sound like much, but it translates to almost one playoff team less per decade for NL Central teams compared with AL West teams.

It's not just playoff pride that the NL Central loses. We also need to consider the financial stakes. To make things simple, let's only consider the money playoff ticket sales make. A playoff team will play anywhere between 3 and 19 extra games. In a typical year, a playoff team might play 8 extra games, 4 of which would be home games. At an average of $30 per ticket and 45,000 tickets per game, that would bring in

($30 per ticket) · (45,000 tickets/game) · (4 games) = $5,400,000.

That extra $5 million goes a long way toward bringing in another starting pitcher to help you make the playoffs again next year, and we haven't even discussed the added revenue from TV advertising and merchandising. Clearly, there's a competitive imbalance. Fortunately, there's hope on the horizon for NL Central fans. As I write this, there are plans to move the Houston Astros to the AL West starting in 2013.

There's No Closet in Football

In 2007, little-known former NBA center John Amaechi made headlines with his autobiography *Man in the Middle,* in which he came out publicly as a gay man. This sparked much debate about the problems gay and lesbian athletes face and the attitudes straight athletes have about homosexuality in the locker room. At one point, I overheard someone say, "Whatever, gay guys don't play football!" **What's the probability that at least one NFL Hall of Famer is gay? How many Hall of Famers (HOFs) should we expect to be gay?**

As I write this, there are currently 259 individuals enshrined in Canton at the Pro Football Hall of Fame. A web search shows a wide range of statistics on homosexuality. In the 1950s, Alfred Kinsey's research on human sexuality reported that 37 percent of men had at least one homosexual experience. Generally, estimates for the percentage of men who are gay range from 1 to 20 percent. In the 2008 US presidential election, exit polling showed 4 percent of people identifying as gay, lesbian, or bisexual.[110] Taking this 4 percent figure, we might estimate there are

$$\text{\# gay HOFs} = (\text{\# of HOFs}) \cdot (\text{fraction of men who are gay})$$
$$= (259 \text{ HOFs}) \cdot (0.04 \text{ gay men for every straight man})$$
$$= 10 \text{ gay HOFs.}$$

We should expect that about ten NFL HOFs are gay, but what is the probability there is at least one gay Hall of Famer? If 4 percent of men are gay, then the probability of a man being straight must be 96 percent. From this, we know that the probability that all NFL HOFs are straight is just $0.96^{259} = 0.003$

---

110 This figure presumably refers only to those members of the gay community who were "out" at the time of the poll. The actual number may be significantly larger.

percent. This means the probability of there being at least one gay player in the NFL HOF is

$$1 - 0.96^{259} = 99.997 \text{ percent.}$$

There is only about a 3 in 100,000 chance there are no gay NFL HOFs.[111] Sadly, given the negative opinions many still have about homosexuality, it may be quite some time before a still-active athlete comes out of the closet to become the gay Jackie Robinson. On the bright side, however, we still have a few years before the inevitable FavreRetirementVickDogFighting-TigerWoodsAffairHaving-esque media circus ESPN is sure to manufacture when it happens.

---

111 Those who have studied probability know that I've made a very big assumption in computing this probability. We've assumed the probability of being gay is uncorrelated with the probability of being an NFL Hall of Famer. This is not necessarily the case. For instance, if football players are perceived as hostile toward gays, it would be less likely that a gay person would choose to take up the sport. If this were the case, then the two probabilities would be dependent on each other, and we couldn't simply multiply the numbers together as we did above. This is an example of conditional probability. We'll talk more about this in the "Trading Cards" problem.

On Kevin Bacon and Hank Aaron

You're on a plane sitting next an attractive blond. Struck by her beauty, you decide to chat her up. Over the course of the conversation, you discover that she once watched her cousin play in a Little League game with your sister's new boyfriend. The heavens must be sending you some sort of a signal! This is clearly the woman of your dreams!

Chance events often create the illusion that something deep and mystical is in the works. In reality, random coincidences like the one above should be expected based on probability. Given the large number of possible coincidences that could happen, it would be very unlikely if none of them did. This is one of the principles behind the Six Degrees of Kevin Bacon game or its baseball variant, Six Degrees of Hank Aaron.[112]

In Six Degrees of Hank Aaron, you're given a random baseball player and you have to find a series of players that leads to Hank Aaron. The number of steps in the series is called the "Aaron number." For example, Yogi Berra has an Aaron Number of two because he played on the 1965 Mets team with Warren Spahn who played on the 1954 Braves team with Hank Aaron.

You should almost always be able to find a path to Hank Aaron with no more than six players. In truth, there's nothing particularly special about Hank Aaron. Players change teams frequently, so the chances are good that a player played with somebody who played with somebody who played with somebody who. . . . Just about any player can be linked to any other player in six degrees or less. If you doubt this, consider me as an example. **What's my Aaron number?**

I'm stretching the definition Aaron number a little bit here, as the game is

---

112 Scientists have their own version of the game, in which they try to link themselves to mathematician Paul Erdös through authors with whom they've published papers. My Erdös number is three.

usually played with only Major Leaguers, of which I am certainly not. If we bend the rules to include nonprofessional teams, then I'm included in the game because I played on a college team that featured Jason Szuminski, the only MIT graduate to reach the Major Leagues.[113] Szuminski's career was short lived, consisting of only seven games for the Padres in 2004, during which he compiled a 7.20 ERA. Though his career was short, Szuminski did have David Wells as a teammate. In 1987, Wells played on a Blue Jays team that had a 48-year-old Phil Niekro on the staff. In 1963, Niekro played on a Braves team that featured one Henry Aaron, leaving me with an Aaron number of four. Let me emphasize: I am in no way comparing my athletic ability to that of any of these Major Leaguers, but the mathematical truth remains. Even a schlub like me can be connected to Hank Aaron in six degrees or less. If you've ever played on a sports team, then by pure chance you can almost certainly trace a connection between you and some famous athlete.

---

113 Perhaps "played" is too strong of a word. I was on the team, but I never so much as pinch ran.

Dream Catchers

It's every baseball-loving kid's dream to catch a home run in the stands. In fact, I'm willing to bet a large percentage of 20-, 30-, and 40-year old dreams involve catching a home run in the stands. Sadly, custom dictates that the few old fogeys who are lucky enough to catch a ball have to relinquish it to the child sitting closest to them. This is especially sad when you consider that one can spend his whole life going to games trying to catch a ball only to lose it to some freeloading jackass 8-year-old who's been spilling mustard and sneezing on you since the third inning. I propose kids should only be allowed to keep balls they themselves catch. **How many games would a kid have to go to before catching a home run?**

You're certainly not going to catch a homerun if you're sitting behind home plate. Most home runs are hit just behind the outfield wall, so you'll only want to buy tickets in this area. Assuming homeruns are all hit within the first 15 rows and that you've bought your tickets accordingly, it's straightforward to calculate the probability of catching a home run.

A baseball diamond is, well, diamond shaped. The lengths of the diamond sides are between 300 and 400 feet, meaning that length of the outfield wall will be about 700 feet. If each chair takes up 2 feet, there will be about 350 chairs per row, or 15×350 = 5,250 seats in which a homerun could fall. If there is a home run, you've got a probability

$$P_{caught} = \frac{1}{5250}$$

of catching it, and a probability

$$P_{not\ caught} = \frac{5249}{5250}$$

of not catching it.[114]

Teams generally hit about 150 home runs per year. Because there are two teams playing in each game and about as many games as there are home runs, there will be about two home runs per game on average. The probability of catching no home runs after N games can be written as

$$\left(P_{not\ caught}\right)^{2N}$$

because you would have witnessed $2N$ home runs and missed all of them. The probability of catching at least one of these balls is then

$$P_{at\ least\ 1}\left(N\right) = 1 - \left(P_{not\ caught}\right)^{2N}$$

With this functional form, you can see that the probability of catching a ball goes up with the number of games attended. To find the number of games you must attend to have a 50/50 shot of catching a ball, we must set $P_{at\ least\ 1}(N)$ equal to 0.5 and solve for N:

$$N = \frac{\log\left(1 - P_{at\ least\ 1}\left(N\right)\right)}{2\log\left(P_{not\ caught}\right)} = \frac{\log\left(0.5\right)}{2\log\left(\dfrac{5249}{5250}\right)} = 3600$$

114 I'm assuming the seats in the home run section are equally weighted (i.e., home runs are equally likely to land on any seat). "But," you interject, "that's not right! Some seats are more likely to have a home run hit there!" True. We could take into account the statistical likelihood of home runs arriving at different seats, but (1) that would be difficult, and (2) as a physicist friend of mine put it, "That's not how you estimate it, that's how you calculate it." To be sure, we've already applied a simple weighting scheme by noting that there are some seats at which you will never catch a home run. This approximate weighting scheme should be good enough to make our results correct to within an order of magnitude.

A kid would have to attend 3,600 games to have a 50 percent shot at catching a ball. Putting this into perspective, he'd need to attend 45 years worth of home games, at which point he clearly would not be a kid anymore. Some might say my "No Giving Kids Balls" policy is selfish, but tell that to the guy who just caught his first homerun after 45 years of trying.

Chess Boxing

Perhaps we've gone too far with the whole biathlon thing. I can understand the original biathlon, in which Norwegian soldiers would train by simultaneously shooting and skiing, as that was at least functional for military members. But indoor cycling gymnastics? Chess boxing?[115] Is there no originality left? Honestly, it's as if someone scrawled a bunch of different sport names onto Post-its and then chose combinations by throwing them against a wall and seeing which ones stuck.

---

115 If you don't believe these sports exist, then I urge you to look them up on YouTube. Don't say I didn't warn you.

On second thought, maybe this isn't such a bad idea. I mean, chess boxing sounds *awesome*. It's a battle of mind versus muscle. Think about it: you could pit Garry Kasparov against Mike Tyson. Could Kasparov checkmate Tyson before the first round ends, and if not, can he withstand Tyson's fury of punches? Will the blows to Kasparov's head diminish his chess-playing ability in the second round? Seriously, how can we make this happen?! **More importantly, how many different "-athlon" combinations could we make?**

To solve this, we need to estimate how many sports there are in total. This is not a particularly well-defined quantity, because anyone can always come up with a new sport. Moreover, even if we only count the sports that have been played up to this point in time, listing every one would be difficult. We could take the list of Olympic sports, but this is far from exhaustive. To simplify the problem, I'm going to consider only the sports listed on Wikipedia. By my estimate, there are well over 1,000 sports listed in Wikipedia's "Outline of sports" page. Assuming 2,000 sports, we can easily calculate the number of "-athlons" in the following way. To begin, we must choose the first sport out of 2,000 choices. Because we can't use the same sport twice, we must choose the second sport out of 1,999 options. Similarly, we must choose the third, fourth, and fifth choices out of 1,998, 1,997, and 1,996 choices, respectively. To find the total number of combinations possible, we must multiply the number of choices for each event. For the biathlon, this gives

$$N_{biathlons} = 2000 \cdot 1999 \approx 4 \times 10^6$$

whereas the total number of triathlons would be

$$N_{triathlons} = 2000 \cdot 1999 \cdot 1998 \approx 8 \times 10^9$$

You likely see the pattern emerging. We're just multiplying a sequence of numbers together. For sports with more events, writing down all the terms gets harder. Fortunately, mathematicians invented a handy notation. You use an exclamation point to represent a multiplication by a series of numbers. For example, 4! (pronounced "four factorial") is equivalent to 1×2×3×4 = 24. Likewise, $N!$ means

$$N! = 1 \times 2 \times 3 \times ... \times (N-1) \times N$$

Using this notation, we can express the number of possible combinations of M sports as

$$N_{sports}(M) = \frac{2000!}{(2000-M)!}$$

Here, we divide by $(2000-M)!$, so that the series of multiplied numbers includes only the largest M terms.

In this way, the number of possible decathlons is expressed simply as

$$N_{decathlons} = \frac{2000!}{(2000-10)!} = 1.0 \times 10^{15}$$

That's one quadrillion different decathlon combinations. It helps to put this number in perspective. If you spent your entire life inventing a thousand new decathlon ideas every second, you'd only have listed 0.1 percent of the total decathlons possible. It's safe to say we're in no danger of running out.

The Most Overrated Record in All Sports

Of all the asinine things ESPN's insufferable talking heads have stated, one sticks in my craw more than all the others combined: Joe DiMaggio's hitting streak is the most unbreakable record in sports.

Any mathematician will tell you that hot hands and hitting streaks are largely the result of statistical fluctuations. I'll get to the mathematical argument in a moment, but for those who remain skeptical that math can debunk the hitting streak myth, let me present three purely nonmathematical arguments for why the talking heads' claim is complete bubkes. First off, DiMaggio wasn't even the best hitter in the league that year. That honor belonged to Ted Williams, who hit .406.[116] Second, have you seen some of the other records? Will White threw 680 innings in 1879. The last pitcher to sniff even 300 innings was Steve Carlton over 30 years ago. What about Cy Young's 511 career wins? C. C. Sabathia may be the active pitcher who's got the best shot at breaking Young's record, but does anyone honestly believe the rotund Sabathia is going to win 20 games a year until he's 50? Third, it's just not that difficult to have a long hitting streak. After all, Ken Landreaux hit in 31 straight games. "Who is Ken Landreaux?" you ask. I rest my case.

What's the probability that some hitter in baseball history would have at least a 56-game hitting streak?

(WARNING: Before diving in, I should tell you that this is unequivocally the longest, most difficult problem in the book. So strap yourself in, because this is about to get bumpier than the acne on Barry Bonds's steroid-injected back.)

---

[116] In the interest of full disclosure, I should state that I am a huge Red Sox fan and I take delight whenever I can piss on something a Yankee does. That said, DiMaggio was a great player even by Hall of Fame standards. The simple fact is that, of his many accomplishments, the hitting streak isn't even the most impressive. For Christ's sake, the guy bedded Marilyn Monroe. If that isn't more impressive than some random statistical aberration, I don't know what is.

This is a pretty difficult problem because the game has changed much in the more than 100 years of history. To start, I'll assume a typical player has a batting average of $P = 0.270$ and on average he gets $A = 4$ at bats per game. Under this assumption, the probability of not getting a hit in any given at bat is $1-P = 0.730$. From this, the probability of having no hits in a game is

$$P_{no\ hit} = (1-P)^A = 0.284$$

The probability P_{hit} of getting at least one hit in a game is one minus $P_{no\ hit}$,

$$P_{hit} = 1-(1-P)^A = 0.716$$

From these two probabilities, we can compute the probability of starting a hitting streak that will last exactly N games. One can find this by noting that the hitter must get a hit in N straight games but must not get a hit in the $(N+1)^{th}$ game. The probability of getting a hit in the first N games is P_{hit}^N, and the probability of not having a hit in the last game is $P_{no\ hit}$. By multiplying these two probabilities together, we find the probability of starting a hitting streak that will last exactly N games,

$$P_{streak}(N) = P_{hit}^N P_{no\ hit}$$

We don't have to worry about multiple ways of arranging the hitting streak because order matters; it clearly would not be considered a hitting streak if the hitless game came somewhere in the middle.

A hitting streak of length N has a probability $P_{streak}(N)$. From this, we can find the average length of a hitting streak,[117]

---

117 Where are my manners? Here I am busting out weird notation without giving an explanation. My apologies. The symbol "Σ" (a Greek letter pronounced "sigma") is shorthand used to represent sums. The numbers on the top and bottom tell you where to start and where to finish adding. For example, we could add all the numbers from 1 to 5 by writing

$$\sum_{i=1}^{5} i = 1+2+3+4+5$$

In general, we can write the sum of any function f(i) from i =a to i = b as

$$\sum_{i=a}^{b} f(i)$$

$$\langle N \rangle = \sum_{i=0}^{\infty} i P_{streak}(i) \approx 2.5 \text{ games}$$

We can now ask the following question: What's the probability that any particular game is part of an N-game hitting streak? Twenty-game hitting streaks are much less likely than ten-game hitting streaks, but there are twice as many games in a twenty-game streak than there are in a ten-game streak. For this reason, if we want to compute the probability that a particular game is part of an N-game hitting streak, we'll need to weight the probability by a factor of N because there are that many games in the streak. First, let me define a quantity Z as[118]

$$Z = \sum_{i=0}^{\infty} (i+1) P_{streak}(i)$$

This Z can be used to weight each of the possible lengths for a hitting streak. I'm including the "+1" to account for the last game of the streak where the player has no hits. The fraction of a player's games in which he has an N-game hitting streak is

$$F(N) = \begin{cases} P_{no\ hit} & \text{if } N = 0 \\ \dfrac{N P_{streak}(N)}{Z} & \text{if } N \neq 0 \end{cases}$$

The fractions of a player's games in which he has at least a 56-game hitting streak is then

$$F_{streak > 56} = \sum_{i=56}^{\infty} F(i) = 1.2 \times 10^{-7}$$

---

[118] Those with a background in physics may recognize this Z as something vaguely similar to the partition function that appears in many statistical physics problems.

There are nine starting players per team, about 20 teams (on average in MLB history), and about 150 games per player per year. This means that over the last 140 years there have been about 3.8 million games. Using , we find that the probability of someone in Major League history having a hitting streak of at least 56 games is

$$1-(1-F_{streak>56})^{3,800,000} = 38\%$$

The result clearly depends on what value one chooses as a typical batting average. If you assume batting averages of 0.280 and 0.260, you get 77 percent and 13 percent, respectively. All these numbers suggest that a 56-game hitting streak is well within the limits of the statistical fluctuations one would expect by chance. In short, DiMaggio was an extraordinarily talented player, but his streak had nothing to do with talent.

Seriously, It's the Most Overrated Record in All Sports

I may be beating a dead horse at this point, but the point needs to be emphasized. Random probability rather than athletic prowess can fully explain hot hands and hitting streaks. To illustrate the point that hitting streaks—even 56-game streaks—should be expected given enough players and time, consider the following question. **What's the probability that Mario Mendoza would hit in 56 straight games?**

Although he was a slick fielder, Mendoza's name has become synonymous with hitting futility. The "Mendoza Line" is a popular term describing any player's batting average that hovers around .200.[119] But even Mendoza would have had a 56-game hitting streak if he played long enough.

In a typical game, a player gets about four official at bats. The probability that a .200 hitter would *not* get a hit in an at bat is 1 − 0.200 = 0.800. The probability that a .200 hitter would not get a hit in four straight at bats is then

$$P_{no\ hit} = 0.800^4 = 0.410$$

From this, we can deduce the probability that a .200 hitter would have at least one hit in four at bats is 1 − 0.410 = 0.590. This is roughly the probability that Mendoza would get at least one hit in a game. From this, we can compute the probability that he would get a hit in 56 straight games:

$$P_{streak} = 0.590^{56} = 1.5 \times 10^{-13}$$

---

[119] Mendoza's career average was actually .215.

As you can see, there is a small but finite chance that even Mario Mendoza would have a 56-game hitting streak. In fact, if Mendoza played two trillion seasons, he'd have a better than 50 percent chance that one of them would contain at least a 56-game hitting streak. If you do the same analysis for DiMaggio's .357 average, you find that for any 56-game stretch, DiMaggio has a 0.003 percent chance of getting at least one hit in every game. Put another way, you'd expect to see at least a 56-game hitting streak about once every 30,000 times. Given that there are about three 56-game stretches over a 162-game season, 30 teams, and nine starters, over the next 100 years you should expect several players to have similar streaks.

I Get No Ref-spect

For one of my new favorite traditions of the playoffs: the one time every spring when the NBA decides it's a good idea to assign Danny Crawford to a Dallas playoff game even though the Mavs are 2–522 when he officiates their games (all numbers approximate).

—Bill Simmons, in an article on ESPN's Page 2[120]

As I write this, the Dallas Mavericks have just beaten the Miami Heat in six games to win the NBA Championship. Throughout the playoffs, it had been alleged that referee Danny Crawford calls a disproportionate number of fouls against the Dallas Mavericks. According to an article posted on ESPN's website,[121]

The Mavs have a 2–16 record in playoff games officiated by Crawford, including 16 losses in the last 17 games. Dallas is 48–41 in the rest of their playoff games during the ownership tenure of Mark Cuban.

Although that seems bad, we're purposefully selecting the worst record by any referee against any team, and it's possible this is just a statistical fluctuation. **What's the probability a record this bad would happen by random chance?**

In principle, the Mavs have about a 50/50 shot of winning or losing any playoff game. In this case, they've had 2 successes out of 18 total attempts. The probability of winning x out of N games can be found from the binomial distribution,

$$B(N, x) = \binom{N}{x} P^x (1 - P)^{N-x}$$

120 Bill Simmons, "NBA Playoffs are 'Wired': Part 1," ESPN, retrieved June 15, 2011, www.sports.espn.go.com/espn/page2/story?page=simmons/part1/110503&sportCat=nba.

121 Tim McMahon, "Danny Crawford to Officiate Game 2," ESPN, retrieved June 15, 2011, www.sports.espn.go.com/dallas/nba/news/story?id=6388692.

where P is the probability of winning any given game, and the strange symbol in the right-hand side of the equation is defined as

$$\binom{N}{x} = \frac{N!}{x!(N-x)!}$$

The symbol above may look complicated, but it has a very simple interpretation. In this case, it's the number of ways you can get x wins out of N games provided you don't care what order you win them in.

Using the values listed in the table below, we find that the probability of winning exactly 2 out of 18 games is

$$B(18,2) = 0.058\%$$

The probability of winning at most 2 out of 18 games is slightly higher:

$$B(18,0) + B(18,1) + B(18,2) = 0.06\%$$

A probability of 0.06 percent seems pretty unlikely, but we've selectively picked the worst record by a team under a specific referee. There are about 40 referees and 30 teams, giving a total of about 1,200 possible records for teams by a given ref. What's the probability that none of them have a record this bad? There's a 99.94 percent chance that a team's record under a referee is better than the Mavs under Crawford, but the probability that all 1,200 possible referee-team combinations are better is

$$0.9994^{1,200} = 45\%,$$

which means there's a 55 percent chance that at least one referee has a record as bad as Crawford's record against the Mavericks. Random probability dictates that more likely than not, some referee in the NBA will have a record this seemingly bad. Through what is likely no fault of his own, Crawford just happens to be the guy with the bad record.

NAME	SYMBOL	VALUE
PROBABILITY OF WINNING ANY GIVEN GAME	P	0.5
TOTAL NUMBER OF GAMES PLAYED	N	18
NUMBER OF GAMES WON	x	2

Thank You Cards

The amount of useless data your brain stores is amazing. I can't remember useful physics formulas for the life of me, but I can tell you that part-time Red Sox outfielder Kevin Romine batted .272 in 1990. I don't pretend that knowing every Triple Crown baseball stat that occurred between 1988 and 1990 is useful,[122] but I'm convinced that my love of baseball stats helped groom me into the math-loving geek I am today. Unfortunately, there's one math problem I never quite figured out until it was too late. You see, there are gaping holes in my 1988–1990-baseball trivia knowledge. The reason is that I only bought the 50-cent packs of baseball cards, in which you got 15 cards per pack. I never actually purchased the $39.99 complete factory set, so there were always a few players whose stats I hadn't seen. I wish I could say it's because I liked the crappy gum that came in the 50-cent packs, but the truth is that my mathematical mind hadn't yet developed to the point at which I could realize this was a terrible strategy. **How many packs of cards must you buy to obtain the complete set?**

There are 792 cards in the Topps 1988 set. Because there are 15 cards per pack, you need at least 792/15 = 53 packs of cards to collect them all. At 50 cents per pack, this would cost you $26.50. This seems like a good value given that the complete set is worth $39.99, but it's very unlikely that you'll get every card on the first try. Just how unlikely is it?

Well, none of your cards can be doubles, which is not so bad at first. Your first card is guaranteed not to be a double. Your second card has a 791/792 chance of not being a double.[123] Your third card has a 790/792 chance of not being a double. The fourth card . . . continuing this pattern for the whole set, you find that the probability of finding no doubles out of your first 792 cards is given by the expression

---

122 Though I was able to impress my brother-in-law when, after hearing of my bizarre talent, he got out his old baseball card collection and quizzed me on the stats.

123 I'm assuming all cards are equally likely to be found in a pack. I'm also assuming that the numbers of cards that Topps manufactures is effectively infinite, so having several copies of card number 1 doesn't decrease the probability of receiving it in the future.

$$P_{no\ doubles} = \frac{792!}{792^{792}} = 7.7 \times 10^{-343}$$

That's bad. If you bought 53 packs of cards every second from the big bang until now, you still would have a 99.99999999 percent chance that none of them would be a complete set, and that's an understatement.

But how many packs would you have to buy until you had a 50 percent chance of collecting at least one of each card?[124]

Topps baseball cards have an index number on the back. Because there are 792 cards in the set, the probability of not getting card number 1 after purchasing N cards is

$$P_{no\ card\ \#1}(N) = \left(\frac{791}{792}\right)^{N}$$

so the probability of having at least one number-1 card is

$$P_{card\ \#1}(N) = 1 - P_{no\ card\ \#1}(N) = 1 - \left(\frac{791}{792}\right)^{N}$$

The probability is identical for cards number 2, number 3, and so forth. From this we can approximate the probability that you have at least one of each of the 792 cards as[125]

$$P_{every\ card}(N) = P_{card\ \#1}(N)^{792}$$

which can be written as

$$P_{every\ card}(N) = \left(1 - \left(\frac{791}{792}\right)^{N}\right)^{792}$$

Setting this probability equal to 50 percent, we can solve for the number of cards N it takes to have a 50 percent chance of collecting every card in the set:

$$N = \frac{\log\left(1 - \sqrt[792]{0.5}\right)}{\log\left(\frac{791}{792}\right)} = 5600 \text{ cards}$$

---

[125] You might wonder why this is an approximation. After all, I told you earlier that you find the probability that two independent events occur by multiplying their individual probabilities together. It turns out that our probabilities here are not independent, but for now, hold that thought until the end of the problem.

You would need to buy about 5,600 cards, or roughly 370 packs of cards to have a 50 percent shot of owning the complete set. That's $185 worth of 50-cent packs.

Sadly, even breaking even is exceedingly difficult. The probability of getting all the cards for under $40 is 1 in 2×10^{85}, which is vanishingly small. Clearly, it's best to save up and just purchase the factory set.

I've glossed over something important in this problem. Up to the point at which we calculated the probability of having at least one copy of card number 1, all our results were exact. The first approximation appeared when we assumed the probability of having at least one copy of every card was equal to the product of the 792 individual probabilities of having at least one copy of any particular card. Because all cards are equally likely, this would have been equal to $P_{card} \#1(N)^{792}$, but there's a mistake in this logic. Since we know that we have at least one copy of card number 1, we know that at least one card in our set will not be card number 2. This changes the probability that at least one of the cards in the set is card number 2. In math-speak, we assumed that the probability of having at least one copy of card number 1 was statistically independent of having at least one copy of any other card. You can see our result for $P_{every\ card}(N)$ is incorrect by plugging in any value of $N < 792$ and seeing that it predicts a finite probability for having a complete set, even though you'd have less than the required 792 cards. As it turns out, our approximation is fairly decent. You can do a quick simulation to confirm this.[126]

The problem above is an example of what mathematicians call conditional probability. It occurs when the probability of one event depends on the outcome of another. Conditional probabilities are notoriously difficult to use because they're often counterintuitive. Consider the famous Monty Hall example. You've got three doors to choose from. Behind one of the doors is your desired prize, a T206 Honus Wagner card valued at $2.8 million. Behind the other doors are piles of sweaty New York Mets jock straps, which you will have to wash by hand if you don't pick the right door. Let's say you choose door #1. Monty says, "Before showing you what's behind door number 1, let

---

126 I discuss simulations later in the "Trading Deadline" problem.

me show you what's behind another door." No matter which door you choose, he always opens a door with smelly jock straps behind it. He then offers you the chance to switch your door with the other unopened door. Do you switch? At first glance, most people believe it makes no difference because there are two doors, so the chances of winning must be 50/50. However, conditional probability correctly predicts that switching doors actually doubles your odds of winning. I'll leave it as a challenge for you to figure out why, but for now, trust me when I say conditional probability can fool even the best mathematician.

CONGRATULATIONS! YOU'VE MADE IT TO THE PROS

These are the problems you do for honor rather than credit.

—An MIT mathematics professor during a Calculus II lecture

If you've made it this far, you're well on your way to becoming a physics sage. But I should warn you: there are rough mountains ahead and we're fresh out of Sherpas. Over the pass, you'll find the most advanced estimation problems in the book, with topics ranging from computational physics to topology to relativity before finally descending on the strange valley of quantum mechanics. If you think you're ready for the trials that lie ahead, by all means continue. But beware! There's no going back after this.

In "Thank You Cards," I illustrated why buying baseball cards by the pack was a bad strategy if your goal was to collect the entire set. In the end, you'd end up paying a lot more money than you would if you just shelled out $40 for the factory set. I forgot to mention one caveat: this calculation assumed you weren't trading cards with friends. By exchanging duplicates for missing cards, obtaining the entire set is easier.[127] Viewed in a strategic light, trading baseball cards can be a valuable tool for learning about probability, not to mention free-market capitalism. **How many trading partners do you need in order for buying cards by the pack to make financial sense?**

As in the "Thank You Cards" problem, our goal is to compute how much money we expect to spend buying cards by the pack. In this case, we have friends who are collectors as well. If we have ten trading partners, we'll need to obtain ten copies of each card for all of us to have a complete set. In this case, we're trying to solve for the number of friends, which I'll call N_f. As before, the probability P of getting any particular card is 1/792.

On the surface, this is a very difficult problem to obtain an answer for.[128] Fortunately, we physicists live during a very special time in history. We get to be lazy! It's very easy to write a computer code that will simulate the system and give an approximate answer for this question. An algorithm for solving this might take the following form:

---

127 When I was younger, I didn't recognize that this was a good strategy. For some reason, I decided it was good idea to trade rare and expensive cards for Marty Barrett cards of which I already had seven copies. (Barrett was my favorite player.) Nick, if you're reading this, don't think I've forgotten how you swindled me out of that Ken Griffey Jr. rookie card.

128 In truth, it's fairly easy to derive an answer for this and the "Thank You Cards" problem by generalizing the binomial theorem. After purchasing N cards, the probability of having n_1 copies of card number 1, n_2 copies of card number 2, and so forth is given by

$$\frac{N!}{(n_1!) \cdot (n_2!) \cdot (...) \cdot (n_{792}!)} P^{792}$$

The difficulty comes from the fact that you must consider every possible sequence for N cards in which there are at least ten copies of each. For those of you familiar with the summation and product symbols, this can be expressed as

$$P(N_f \text{ complete sets}) = N! P^{792} \sum_{n_2 = N_f}^{N} \sum_{n_2 = N_f}^{N - n_1} ... \sum_{n_{792} = N_f}^{N - n_1 - n_2 - ... n_{791}} \left(\prod_{i=1}^{792} \frac{1}{n_i!} \right)$$

Sadly, even though we can write down the solution, evaluating and coming up with an actual number is extremely difficult.

Step 1: Generate a random integer between 1 and 792 and store this integer in a list.

Step 2: Repeat step 1 until each number has been recorded N_f times.

Step 3: Once each number has been recorded N_f times, count the total number of integers stored in your list. This is roughly the number of cards you and your friends will need to buy before each of you owns a complete set.

You'll probably want to repeat steps 1 through 3 multiple times so you can get an average value for the number of cards you and your friends will need to buy. Once you do this, you can rerun the algorithm with different numbers of friends and then use this to determine the break-even point.

If you write such a code, you'll find that the break-even point is around 50 friends.

A "Better" Stat

Who's a better athlete, Michael Jordan or Babe Ruth? Sports enthusiasts debate conundrums like this all the time, but the question is ill defined. What does it mean to be "better"? To compare the two men, we need to extract data from different eras, different skill sets, and even different sports, making the problem a muddy indiscernible mess, of which there is no correct answer. Even when sports and eras overlap, there's room for debate. Magic versus Bird? Brady versus Manning? Crosby versus Ovechkin? And yet all these difficulties can be erased by posing the question slightly differently: who's a better athlete, Michael Jordan or Mario Mendoza? A group of sports fans would likely have no trouble picking a unanimous answer, even though the problem ostensibly creates the same difficulty as before. Why is there trouble in one case and not the other?

On the surface, the answer is obvious. As humans, most of us have a vague notion of what "better" means. I doubt even my own mother would argue that I was as athletic as Michael Jordan or any other professional athlete. When the discrepancy between players is large, it's easy to arrive at an answer, but when two athletes are close in ability, it's impossible. Our conceptions of words like "better" are inherently fuzzy. Consider a word like "perfect," which at first glance seems fairly straightforward. If you're perfect, you don't make any mistakes. A pitcher throws a perfect game if he pitches all nine innings and there are no hits, walks, errors, or base runners of any kind. But is this really perfect? After all, if the pitcher was really perfect, wouldn't he be able to throw only 27 pitches and get outs on all of them? Even this definition has flaws because each hitter would have made contact with the ball, whereas the perfect pitch would be untouchable. Perhaps we should say that in a truly perfect game, the pitcher strikes out all batters with 81 pitches and no batter makes contact. However, even this definition is dubious. What if the first batter struck out and reaches base on a passed ball? The pitcher goes on to strike out the rest of the hitters on three pitches each. He hasn't made any mistakes, and he's done so for even longer than a normal game would require. Is that a perfect game? By the current definition, it is not, even though the pitcher did nothing wrong. Clearly, definitions for words like "better" and even "perfect" are subject to interpretation. **How do we get around the inherent fuzziness in words like "better" so we can determine the best athlete of all time?**

If we had a more precise definition, we could compare athletes and determine with certainty the best athlete of all time. Therein lies the idea behind statistical metrics, or what some physicists call "order parameters." [129] To get around the fuzziness, sports statisticians, or sabermetricians as they like to be called, attach numbers to certain qualities. These numbers serve as a measuring stick with which you can compare athletes or, in the case of the Bowl Championship Series (BCS), teams. Using statistics to compare athletic performance is certainly not a new concept, because batting averages have been recorded almost from baseball's beginnings. However, recently there's been an explosion in the number and quality of stats used to measure an athlete's ability. I'm not going to go into depth on these new stats because there are already several very good sabermetrics books on the market, and it's not a topic I'm intimately familiar with. Rather, our concern here is with the mathematical process one might use to come up with a stat to compare athletes. Specifically, we're interested in how one might synthesize what some might consider the Holy Grail of sports statistics—a single stat that can compare athletes across different sports and different eras.

One simple method to define the best athlete of all time is consensus. After all, we live in a democracy, so perhaps we should just vote on our favorite athletes and see who gets the most votes. This is a technique ESPN and other various sports pages employ whenever they do a "Best Athletes of All Time" survey. This strikes me as the lazy way out. First off, being largely

---

129 In physics, order parameters are often used to determine when a phase transition occurs. In some cases it's as simple as looking at the density of a substance transitioning from a liquid to a gas, but in many others it's quite complicated. I had some physicist friends who once looked at fluctuations in home run output to see if they could define an order parameter that would quantify the likelihood that a player took steroids.

subjective, the results are often biased. A Celtics fan will almost always rate Kobe lower than a Lakers fan would. Second, majority opinion doesn't always jive with facts. In the United States today, over 50 percent of people don't believe in evolution. If democracy actually determined facts, all the dinosaur bones in all the museums in America would spontaneously combust. Finally, if you look at the results from some of these surveys you'll see why they're pretty dubious. I once saw a survey ranking the greatest Yankees of all time, with Derek Jeter a very close second to Babe Ruth and orders of magnitude ahead of Lou Gehrig, Mickey Mantle, and Joe DiMaggio.[130] Don't get me wrong. Jeter is certainly a great player, but there are times he gets more praise than Jesus. Frankly, given Jeter's abysmal range factor, I suspect Jesus would have been a better fielder even with the stigmata.[131]

If you're a die-hard Jeter fan, then perhaps you require more proof that consensus in ranking athletes has shortcomings. In 1999, ESPN assembled a panel of sports journalists to vote on the Top 50 Athletes of the Twentieth Century, and they ranked Secretariat as the 35th best athlete. Forgive me, but I can't help thinking Secretariat should have been ranked somewhat higher given that he (A) won the triple crown, (B) set race records at both the Kentucky Derby and the Belmont Stakes, and (C) was a $@#-ing horse! Call me crazy, but I suspect a rigorous statistical analysis may be a little more accurate than the whims of opinion.

Perhaps asking for a single stat that determines overall athletic success is asking too much. Maybe we should use a collection of statistics to determine who the best athlete is. Unfortunately, even this is fraught with potential problems. Let's say we're comparing Red Sox leftfielders from different eras using only the Triple Crown categories, with the rule that if a player wins in two out of the three categories, then he is considered a better player. Let's take Ted Williams's 1946 season, Carl Yastrzemski's 1967 season, and Jim Rice's 1983

---

130 Jeter's best batting average in a season is only nine points ahead of Gehrig's career batting average. When you take into account the power numbers, Jeter's best offensive season would have been terrible by Gehrig standards.

131 Admittedly, I'm probably going to hell for that last line, but whenever I think about playing catch with Jesus, I can't help imagining the same thing: you throw him the ball, it passes through his hand, it rolls to the backstop, he has to go get it, you throw him the ball again, it passes through his hand, it rolls to the backstop. . . . On a related note, complaint letters and hate mail can be sent to Aaron Santos, c/o Running Press, 2300 Chestnut Street, Suite 200, Philadelphia, PA 19103.

season. With a line of .342/38/123, Williams beats Yaz's .326/44/121, but loses to Rice's .305/39/126. Does Rice win? No, because he still loses to Yaz. This is sometimes called the "rock-paper-scissors" problem. Even if A beats B and B beats C, there's still a possibility that C beats A, so you can't define a nice linear hierarchy. Much like Triple Crown categories, real life has multiple variables, so it, too, cannot have a clear hierarchy. As such, defining perfect metric to measure overall athletic talent is almost certainly impossible, but we've already come this far, so let's see what happens if we try to weight all of these different stats and compress them into a single number.

Let's say you've got a definition for your metric. Using this metric, you can calculate a single number that will rate athletic performance across a host of disciplines. How do you know if it works? Fundamentally, we're attempting to define a number, which, when we calculate it, gives a quantitative measure of something we already understand intuitively. The number is only useful when comparing athletes whose abilities are so close that our intuition can't resolve who's better. In order for our metric to work, a minimum requirement is that it correctly predicts what our intuition says for all the clear-cut Mendoza versus Jordan cases. If it works for these, then maybe it will work when comparing Ruth and Jordan. By testing all these simple cases, you can get a good idea whether or not your ranking system works.

There's one small catch. Let's say you do find a metric that gives intuitively correct results for all the easy cases. Is that the only metric you could have come up with? There's a problem of uniqueness here, and it's very likely that two different metrics can predict all the same clear-cut cases but differ in results when athletes are close. Depending on how you weight different factors, you could easily say either Jordan or Ruth was better. The inability to define a unique solution plagues problems of this type. There is inherently some arbitrariness in the "better athlete" definition, and unless you can get everyone to agree on what constitutes a good metric, then I'm afraid you're out of luck. I suspect we'll continue to debate "best athlete" questions for the foreseeable future.[132]

---

132 If you like thinking about problems like this—and even more so if you like thinking about the problems in the probability section—you might want to explore more on actual sabermetrics. Unlike ethereal definitions of "better," sabermetricians develop metrics to measure specific abilities and how they relate to overall team performance. Given the money that teams are pouring into sabermetrics, it appears this burgeoning field will be around a long time. I suspect it's a good fit for any mathematically inclined sports fan.

Rodeo Clowns

A couple years back, my in-laws took me to a rodeo. The rodeo was . . . um . . . er . . . that is to say. . . . Listen, I'm going to be honest. I don't really know anything rodeo except that it involves rope. Here's a nice rope topology riddle you can do:

You'll need two people and two pieces of rope (or some other rope-like material). Make each rope piece about four feet in length. Tie the first rope around person A's hands to make a set of rope handcuffs. A's arms will now form a circular loop with the rope. Do the same for person B, but make B's circular loop interlock with A's. **Without cutting the rope, untying it, or sliding it off the hands, how can you unlink persons A and B?**

This is a great trick to break out at parties, especially if there's a particular lad or lass with whom you'd like to be tied up. Give your victims some time to ponder the conundrum. Undoubtedly, they will try twisting themselves into a wide array of Cirque du Soleil-esque contortions—which outside observers will find quite hilarious—before eventually giving up and concluding that the feat is surely impossible. At this point, it's your call whether you want to step in and relieve the players of the mystery or make them suffer longer by assuring them the trick can be done. I recommend the latter until such time as the victims begin strongly considering beating you to a bloody pulp, at which point it's clearly time to step in with all your brilliance and solve the riddle.

Grab the middle of B's rope and pull it next to one of the loops on A's wrists. Pull B's rope up through the loop on the inner part of A's wrist, then carry it over A's hand and out the other side of the loop on the back side of A's wrist. Your subjects will now be untangled.

Admittedly, this may seem like an odd problem to include in a book of approximations. "Where are the numbers?" you ask. Although there aren't numbers in this problem, your brain is still making approximations. The reason that so many people have difficulty solving this riddle is that most of us assume the handcuffs have the same topological structure as interlocking rings, which cannot be separated. The rope handcuffs don't have this structure, of course, or the riddle couldn't be solved. It's interesting that even after being shown the solution, people often have a hard time "seeing it," perhaps because the assumption of interlocking rings is so strongly ingrained. This goes to show that, whether we want to or not, our brains constantly make assumptions about the world we live in, and we would do well to recognize when the assumptions we make affect the conclusions we draw.

Einstein at the Bat

In 2010, Detroit Tigers pitcher Armando Gallaraga was one out away from a perfect game. Jason Donald, the 27th batter, grounded a ball to first baseman Miguel Cabrera, who flipped to Gallaraga covering. Umpire Jim Joyce called the batter safe despite the fact that replays clearly showed Donald was out. Perhaps Joyce was not necessarily wrong. **Is there any reference frame in which Joyce's call would have been correct?**

You: Replay shows Donald was out by about 0.1 seconds. How can that be wrong? And what is this mysterious "reference frame" of which you speak?

Me: To a nonphysicist, this question may sound crazy, but hear me out. In 1905, Albert Einstein came up with his special theory of relativity. In addition to the famous $E=mc^2$ equation that predicted a future with atomic bombs and nuclear power, the theory also predicted that events that seem to happen one after another do not necessarily happen in that order. For example, events that seem to occur simultaneously will only appear that way when you're traveling at a certain speed (what we physicists call a "reference frame"). For an observer moving at a different speed, the events might occur in a different order.[133]

You: So you're saying I can be born before my parents?

Me: No. Logically that can't happen because those two events are causally related: your parents have to have been born before you can be born. But once you *are* born, you can age faster than they do and ultimately grow older than them by traveling really, really fast (i.e., close to the speed of light).

You: Trippy.

Me: Exactly. Anyway, the only way two events are causally linked (i.e., the only way one event has to happen before the other) is if the light traveling from the first event arrives at the location of the second event before the second event happens. In Gallaraga's case, if the light traveling from the "ball hitting his glove" event doesn't arrive at the first-base bag before the "Donald reaching first base" event, then those events could have happened in a different order.

You: Can I buy pot from you?

Me: Physics is all the pot you need, my friend. To find out if there's any reference frame in which Joyce made the right call, all we have to do is find

---

133 There may be those readers among you who scoff at this and the other predictions of relativity and say, "Well, that's just a theory!" I applaud you for your skepticism, which is indeed a valuable skill to develop, but I invite you to read up on experiments done to test relativity. You'll find that there has not been a violation to date. On a related note, if you are in the "just a theory" camp, please remember that saying something is "just a theory" is the logical equivalent of saying, "We're only 99.9999999 percent certain it's true."

out how long it took the light leaving the "ball hitting Gallaraga's glove" event to reach the bag. Because Gallaraga was stretching for the ball, we might assume the distance from his hand to the first-base bag was about 2.5 meters. As the speed of light is 3.0×10^8 m/s, we can derive the time it took for the light to reach the first-base bag:

$$\Delta t = \frac{\Delta x}{v} = \frac{(2.5 \text{ m})}{\left(3.0 \times 10^8 \text{ m/s}\right)} = 8.3 \times 10^{-9} \text{ s}$$

It would take about 10 nanoseconds for the light to reach the bag. That's a tiny amount. Donald was out by about 0.1 seconds, which is huge compared to that. In fact, even if the play was so close that human eyes could barely distinguish which event happened first, there's still no way relativity could account for the blown call.[134]

You: So . . . you're just wasting our time with this one.

Me: Hey, you learned some relativity, didn't you?

---

134 This would be about 0.03 seconds. We know this because you can't see a television flicker, and the screen updates about once every 0.03 seconds.

Relativistic Marathons

I've often heard that running can help you live up to five years longer. Provided a Mack truck doesn't hit you during your jog, this is almost certainly true due to the increased cardiovascular health that results from this activity. However, even if one neglects the added health benefits, it's still true that you would live relatively longer than your sedentary brethren. According to Einstein's theory of special relativity, objects in motion relative to some inertial reference frame will age slower than objects at rest when time is measured in the same frame. Put simply: you age more slowly while running around. **If you run marathons, how much longer than a sedentary person will you live because of relativistic effects?**

A marathon is 42 kilometers (~26 miles) in length. A reasonably good runner can complete one in three hours by traveling at a speed of 3.9 m/s. By Einstein's theory of relativity, we know that a moving object ages slower than one at rest. The time t_{moving} that passes for a moving object is given by

$$t_{moving} = t_{not\ moving}\sqrt{1 - \frac{v^2}{c^2}}$$

where $t_{not\ moving}$ is time that passes for objects at rest, v is the speed of the moving object, and c is the speed of light. Using the values above, we can calculate the percent difference between the time that passes in the moving frame to that of the nonmoving frame:

$$\frac{t_{not\ moving} - t_{moving}}{t_{not\ moving}} = \left(1 - \sqrt{1 - \frac{(2.3\ \text{m/s})^2}{(3.0\times10^8\ \text{m/s})^2}}\right) \approx 2.9\times10^{-15}\%$$

That's a tiny percentage difference. To put that in perspective, if you ran a three-hour marathon every day for 80 years, you would age about 26 nanoseconds less than someone who remained sedentary. If you wanted to age 5 years less, you would need to run 99.9 percent the speed of light, or roughly 670 million mph.[135]

135 Many subtleties exist in regard to this problem, including what happens for noninertial frames (i.e., frames that are accelerating). These subtleties contain some rich physics, but sadly, it's beyond the scope of this book. My hope is that these advanced problems will inspire you to seek answers to these and other questions in other physics literature or classes.

It's Not Easy Being 7-foot-1 and Green

In "The Pressure's On," I showed why NBA big men like Shaq are inherently prone to knee problems. Sadly, Einstein's theory of relativity dictates that this is not the only misfortune that will befall the Big Diesel.

After Einstein came up with special relativity, he generalized the theory to describe the effects of gravity. The theory predicts that large masses can distort the shape of space-time. For example, the flight of a baseball might curve downward more if there was a mass beneath the stadium large enough to bend space. If you've been following the book so far, you're probably not too surprised because you know that a larger mass would produce a gravitational force that would pull on the object. In general relativity, Einstein showed that gravity is more accurately described by a bending of space than it is by the force we discussed earlier. Although you might think this is crazy talk, it turns out that all experimental tests so far have confirmed the theory. In fact, the theory is so well established that even engineers employ it: the GPS in your car requires the use of general relativity to predict your location with such high precision. One side effect of the theory is that objects closer to a large mass like the Earth will age more slowly. **Over the course of a lifetime, how much more will Shaq age than the 5-foot-3 Mugsy Bogues?**

According to general relativity, time passes more slowly if you are at a lower gravitational potential (i.e., closer to the Earth). The effect is small unless you're talking about a very large mass like a black hole, but even for the Earth the effect is measureable. In a recent experiment, scientists at the National Institute of Standards and Technology measured a time difference using high precision atomic clocks that differed in altitude by as little as one-third of a meter.[136] The time that passes for an object in a gravitational field is given by

---

[136] John Matson, "How Time Flies: Ultraprecise Clock Rates Vary with Tiny Differences in Speed and Elevation," Scientific American, http://www.scientificamerican.com/article.cfm?id=time-dilation.

$$t_{field} = t_{no\ field}\sqrt{1 - \frac{2GM}{rc^2}} \approx t_{no\ field}\left(1 - \frac{GM}{rc^2}\right)$$

where $t_{no\ field}$ is the time that would have passed in the absence of any field, G is the fundamental gravitational constant, M is the mass of the large body creating the gravitational field, r is the distance between the object in question and the center of mass of the large body, and c is the speed of light.[137] Because Shaq is taller than Bogues, he is further away from the Earth and thus time will pass more quickly for him. The difference in time that passes for each man can be written as

$$\Delta t = t_{Shaq} - t_{Mugsy}$$

$$\approx t_{no\ field}\left[\left(1 - \frac{GM}{r_{Shaq}c^2}\right) - \left(1 - \frac{GM}{r_{Mugsy}c^2}\right)\right]$$

$$= \frac{GMt_{no\ field}}{c^2}\left[\left(\frac{1}{r_{Mugsy}} - \frac{1}{r_{Shaq}}\right)\right]$$

$$= \frac{GMt_{no\ field}}{c^2}\left(\frac{r_{Shaq} - r_{Mugsy}}{r_{Shaq}r_{Mugsy}}\right)$$

137 You may wonder where the approximation in the second half of this equation comes from. All I can say is, "Hey, look kids! There's Big Ben and there's Parliament." As I mentioned in an earlier footnote, the Taylor series can be used to approximate functions when the variable inside is small. In this case, we're approximating the function describing t_{field} by using the fact that GM/rc^2 is small.

Here, $g = 9.8$ m/s^2 is the acceleration of gravity at the surface of the Earth.

Using the values assumed in the table below, we find $t = 0.15$ microseconds. Over the course of 80 years, Shaq should have aged an extra 0.15 microseconds more than Mugsy. Even more surprising, Shaq's head is almost a microsecond older than his feet!

NAME	SYMBOL	VALUE
FUNDAMENTAL GRAVITATION CONSTANT	G	6.67×10^{-1} N·m^2/kg^2
MASS OF THE EARTH	M	6.0×10^{24} kg
LIFETIME	$t_{no\ field}$	80 years
SPEED OF LIGHT	c	3.0×10^8 m/s
DISTANCE FROM SHAQ TO CENTER OF THE EARTH	r_{Shaq}	6,400 km + 7'1"
DISTANCE FROM MUGSY TO CENTER OF THE EARTH	r_{Mugsy}	6,400 km + 5'3"
ACCELERATION OF GRAVITY AT EARTH'S SURFACE	g	9.8 m/s^2

Quantum Ball

You've entered a strange world. A world so bizarre in scale with phenomena so perplexingly eerie, that I dare not attach any math to it. It is . . . The Quantum World.

From out of the shadows, you hear an unfamiliar voice call out your name. "Who's that? Where am I?" you gasp.

"You've entered the Quantum World. To make it back home, you have to beat the Mechanic at his game."

Just then, you hear the bellowing voice of the announcer declare, to raucous applause, "Now pitching: the Quantum Mechanic." As the crowd settles, you hear the announcer pronounce, "Now batting . . . um . . . uh . . . this guy."

"You're up, kid. Here, you're going to want to swing with this," your new guide says as he hands you a mysterious bat, which seems to float in your hands. As you jostle it, it emits a faint glow.

"This bat is . . . er . . . "

"You're welcome."

Mysterious glowing bat in hand, you stride into the batter's box. Staring down at the Mechanic, you notice the unusual red ball in his hand. Just then, he comes to his set position.

"Here's the windup—and the pitch . . . "

As the Mechanic goes through his motion, the ball transforms, changing colors from red to yellow to green to blue before finally settling on violet just as he releases it.

You swing with all your might, sure you've made contact! But what's this? Something's wrong. The catcher has the ball. He's throwing it back to the pitcher. It's as though the ball went straight through your bat.

"STRIKE ONE!" the announcer exclaims.

"How did I miss that?" you ponder aloud. "Did it curve?"

"There's no air here, son. Curveballs are the least of your worries. But don't worry. That won't happen every time," your new guide reassures you.

You step back into the batter's box, not exactly comforted by your guide's words. Lowering yourself into your stance, you wait for the next pitch.

"The Mechanic comes to his set. Here's the pitch . . ."

As the Mechanic releases the ball, you wince. For even though you can see as clear as the day, there appear to be two balls superbly positioned with one traveling straight over the plate and the other moving at near light speed toward your face. As you quickly dive out of the way, the latter ball vanishes, leaving only the first.

"STRIKE TWO!" the announcer yells, to the thunderous applause of the spectators.

"That's not fair! He can't do that!" you protest.

"Yes, he can," your guide admonishes. "Just because you didn't imagine it would happen this way doesn't mean it breaks any rules. It's all perfectly logical."

"Whose side are you on anyway?" you snap back. Your guide only smiles.

Over the PA, the announcer waxes, "The Mechanic is really throwing some heat today. We better see what he's hitting on the radar gun . . . "

Taking a deep breath, you enter the batter's box once more, certain you'll soon be doomed to spend the rest of your days in this wretched Quantum World.

"Here's the windup and the pitch . . . "

Once again, you swing with all the strength you can muster, but it's no use. You're sure you've missed. And yet something odd happens. The crowd has fallen deafly silent. They're leaving the stadium. You look up to the radar gun which reads $0.999999999999999c$.

"Well, that's the ballgame," the announcer says. "Be sure to join us tomorrow for Quark Bobblehead Night."

"What happened?" you ask your guide.

"Game's over."

"What? Why?"

"Uncertainty principle," he replies. "That ball could be miles from here."

The phrase "uncertainty principle" needs a moment to settle before you can reply.

"I think I've heard of that. It means you can never measure the position and velocity of an object at the same time."

"Your teachers got it wrong, kid. It's not that you can't *measure* an object's exact position and velocity. It's that an object can never *have* an exact position and velocity."

Upon hearing your guide's words, you're suddenly whisked away, back to the familiar macroscopic world, never again to experience life at the quantum level.

HANGING UP YOUR SPIKES

Me: *Hey, I gotta get going. My mom's picking me up.*
Muscleless Wonder: *Fine, just leave if you never want to accomplish anything and want to end up getting stuck in New Bedford for the rest of your life.*
Me: *[After a long pause.] So I'll see you tomorrow, right?*

Yogi Berra once eloquently said, "It ain't over til it's over." I'm afraid my part in this journey is over, but that doesn't mean yours has to be. You can keep training your mathematical mind as long as you want. I mean that. Unlike sports, you can have fun playing math games well into your golden years before you even consider retiring. (I'm looking at you, Brett Favre.) Over a lifetime, a lot of fun and utility can be had out of making estimations. Hopefully this book has showed you how.[138] In any event, good luck and have fun estimating!

---

[138] If not, you may want to reread the book. In fact, you probably want to buy a second copy just in case you accidentally damaged the first one. Ideally, it'd be good to purchase several copies to compare, though if you want to do a scientific study, you'll need at least 100 to get good statistics. You'll probably also want a control group, for which I recommend buying another 100 copies of my first book *How Many Licks? Or, How to Estimate Damn Near Anything.* It may cost a bit, but it's for science, and here, we're nothing if not scientists.

Thank Yous

This book could not have been completed without the help of many remarkable people.

Writing this book would have been impossible without the many years I spent in the weight room discussing sports physics with Norm "The Muscleless Wonder" Meltzer. If I hadn't trained in Norm's strength-and-conditioning program, I surely wouldn't be the successful physicist I am today. Instead, I'd probably be off winning Olympic medals or something.

I'd like to thank Sean Robinson and Laurie Santos for their very helpful recommendations. I'd also like to thank the following people for various suggestions and permissions: Michael Bergman, Jessica Braley, Matthew Brodo, Kurt Collins, Edward Farhi, Ahmie-Woma Farkas, John Fries, Abbie Stählin Gentry, John Eric Goff, Eric Jankowski, Sophie Kapsidis, David Kessler, Kendall Mahn, Alex Newman, Jason Smith, Jason Pacheco, Brian Pothier, Jason Stalnaker, and Carl West.

To my agent Sorche Fairbank, illustrator Mario Zucca, my editors Jennifer Leczkowski, Cisca Schreefel, and Josephine Mariea, and all the wonderful people at Running Press: thank you for all your hard work in putting this book together.

To my family—Mom, Dad, Sue, Laurie, Mark, and my Nebraska family—Uncle Rich, Aunty Mom, Cabello the Terrible, Derek, Angela, and Baby Wack-Lack: thank you for your wonderful support.

Finally, thank you to my wonderful wife, Anna Lackaff, without whom none of this would be possible.

About the Author

Aaron Santos realized his athletic career wouldn't amount to much when his football coaches told him, "Defensive linemen are big and slow. We didn't put you on defensive line because you were big." With sports no longer an option, he decided to pursue a career in physics. At the time of publication, he will either still be an assistant professor in the Physics and Astronomy Department at Oberlin College or looking for employment. In addition to *Ballparking*, Santos authored *How Many Licks? Or, How to Estimate Damn Near Anything*, another humorous work on the art of estimation. When not teaching, writing, or doing science, he occasionally consults with Muscleless Wonder Strength and Conditioning. Although he's given up on making it to the pros, he still harbors dreams of becoming a professional miniature golf course architect.

You can find out more about Aaron at aaronsantos.com, by following him on Twitter @aarontsantos, or by reading his blog at diaryofnumbers.blogspot.com.

INDEX

Aaron, Hank, 168–169

Acceleration

defined/equation, 62, 65

equation of motion and, 63–64, 63n35

Aging and general relativity, 204–206, 206 (table)

Algebra description, 60–61

Ali, Muhammad, 29, 69n39

Amaechi, John, 166

Anderson, Brady, 47

Angular momentum, 122–123

Angular momentum conservation

figure skaters, 124–125

track running direction/day length change, 126–127, 127 (table)

Angular velocity, 123, 124–125, 129n84, 149, 158

Approximations. *See* Estimations

Archimedes, 34, 99

Armstrong, Lance, 24–25

Assumption analysis, 119

Autograph vs. tipping, 29–30

Bacon, Kevin, 168

Barrymore, Drew, 6

Baseball

buying Yankees exhibition game vs. buying all game tickets, 20

estimating catcher's squats, 17

"fireballer" vs. "having a cannon," 150–151, 151 (table)

hitting streak record, 176–179, 180–181

in Quantum World, 207–209

rat droppings in hot dogs, 21

retired numbers problem, 31–32

See also specific individuals

Baseball card packs

probability of acquiring set, 185–188, 191

probability of acquiring set with trading, 191–192, 191n128

Baseball players/steroids

bat speed, 105

fly ball distance, 106, 106n70

reaction time, 104

strength, 103–106

Baseball Simulator, 96–97

Baseballs

curve balls, 149, 157–158, 158 (table)

force friction parameters of, 152–153, 152 (table)

radar gun measurement location, 152–153, 152 (table)

rising fastball, 157–158, 158 (table)

speed drop (pitcher to plate), 152–153, 152 (table)

traveling in "space stadium," 96–97

Basketball

bolts in court, 16

buzzer beaters vs. free throws, 57–59

"on fire" phrase, 146–147

Basketballs

expansion if on fire, 146–147

pressure/inflating, 135, 146

Beard possibilities, 46–47

Beer curls and weight loss, 116–117

Bell, Cool Papa, 69n39

Bench, Johnny, 17

Berra, Yogi, 32n13, 168, 210

"Best athlete" question/problems, 193–196

Bias and estimations, 15

Biathlon combination possibilities, 173–175

Big Show, The (WEEI radio), 24

Blimp, punting/hitting blimp, 81–82

Bogues, Mugsy, 51–52, 204–206

Bolt, Usain
records, 68, 74, 118
vs. Chris Johnson, 76–77
vs. racecar, 74–75

Bonds, Barry, 176

Boobs and harmonic motion, 130–132, 133

Brewster's Millions (movie), 20

Buck, Joe, 17

Buckner, Bill, 44

Bull riding, 94–95

Buoyant force
description/equation, 134
throwing helium-filled football, 142–145, 145 (table)

Cabrera, Miguel, 199

Calculus description, 61

Calorie, 107n71, 110, 111

Campanella, Roy, 5

Carlton, Steve, 176

Cavendish, Henry, 100

Celsius/Kelvin conversion, 135n89

Chamberlain, Wilt, 42–43

Chapman, Aroldis, 152, 153

Chara, Zdeno, 18

Chase, Chevy, 148n98

Cheers (television sitcom), 16

Chess boxing, 173, 174

Chestnut, Joey, 30

Clemens, Roger, 105

Comparing athletes/problems, 193–196

Competitive eating, 109–110

Conditional probability, 188–189

Cosine of angle, 80n45

Costner, Kevin, 23

Coupling and harmonic motion, 132–133

Cowan, Glenn, 137n91

Crawford, Danny, 182–185, 184 (table)

Culture Brain, 96

Curling, 101–102, 101n67, 102n68

Dallas Mavericks, 182–185, 184 (table)

Damon, Johnny, 150

Damping and harmonic motion, 132

Dead or Alive (video game), 130

Dean, Dizzy, 157

"Delta" symbol defined, 64n36

Dempsey, Tom, 56, 56 (table)

Density
description/equation, 134
"rho" symbol and, 93n59, 134
of water/human bodies, 34n16, 41

Derivatives defined, 64

Dickey, Bill, 32n13

DiMaggio, Joe, 7, 32n13, 176, 176n116, 179, 181, 195

Dimensional analysis, 85–87, 98, 118–119, 160, 162

Donald, Jason, 199

Donkey Kong video game, 38

Drag coefficient, 148, 150, 151 (table), 152, 152 (table), 154, 155 (table), 155n102

Drag equation, 154–155

Drag force, 145, 148–149, 148n98, 150, 154–155, 160–161
See also Friction

Drift velocity, 69n41

Duck Hunt (video game) dog, 36–37

Earnhardt, Dale, Jr., 74–75

Egg-drop contest, 140n93

Einstein, Albert, 200, 202, 204

Electricity speed, 69–70, 69n41

Energy
ATP and, 112
changing forms, 107–108
competitive eating, 109–110
elderly breaking bones and, 140n94
human rate of energy use, 111,
111n73
jumping and, 140–141
Michael Phelps and, 111, 111n74
power and, 108
units of, 107n71
work and, 107n71
Energy conservation
description, 107–108
Newton's laws and, 90n55
Energy/weight loss
beer curls, 116–117
bench presses, 114–115
climbing stairs, 112–113
sex, 118–119
English system of units, 9n4
Entertainment Weekly, 22
Erdös, Paul, 168n112
Error ranges, 9n2
ESPN, 5, 77, 103, 167, 176, 182, 194–
195
Estimations
physics problems and, 13
rules for making, 14–15
See also specific examples
"Eta" symbol defined, 148
Exponentiation, 61

Factorials, 175
Farhi, Edward, 48
Fermi, Enrico, 14n5
"Fermi method," 14–15, 14n5
Ferrell, Will, 124
Fever Pitch (movie), 6
Field of Dreams/male tears, 22–23
Figure skaters and angular
momentum conservation, 124–125

Finger, Rollie, 46
Flashlights, "shake-weight" types,
69n40
Flat Earth concept, 86n51
Flatulence effects on weight lifting,
98
Football Hall of Fame, 166–167
Footballs
geometry of, 145n97
punting/hitting blimp, 81–82
throwing helium-filled football, 142–
145, 145 (table)
Forces
athletes' strength increase and, 103–
106
baseball traveling in "space stadium,"
96–97
bull riding example, 94–95
buoyant force, 134
curling example, 101–102, 101n67,
102n68
Newton's laws, 88–91
pulleys needed to lift Earth, 99–100
types, 89n54
units of, 91n57
See also Gravity
Ford, Whitey, 32n13
Formula One racing, 45
Friction
assumption as absent/negligible, 148
baseball curve balls and, 149
baseballs speed drop (pitcher to
plate), 152–153, 152 (table)
drag coefficient, 148, 150, 151 (table),
152, 152 (table), 154, 155 (table),
155n102
drag equation, 154–155
drag force, 145, 148–149, 148n98,
150, 154–155, 160–161
equation for approximating, 148–149
"fireballer" vs. "having a cannon,"
150–151, 151 (table)
radar gun measurement location,
152–153, 152 (table)

sumo wrestler's fat/stopping bullet, 159–162, 162 (table)

swimming with cement loafers, 154–156, 155 (table)

Functions/functional dependencies

notation, 48n28

overview, 48–49

"Fuzzy" numbers, 8–9, 8n1, 9n2

Gallaraga, Armando, 199–201

Gates, Bill, 7

Gateway Arch, Saint Louis and pole vaulting, 120–121

Gay NFL Hall of Famer probability, 166–167, 167n111

Geek humor, 6

Gehrig, Lou, 32n13, 195, 195n130

General relativity, 204–206, 206 (table)

Gilbert, Jarron, 142

Golfing in solar system, 55–56, 56 (table)

Gonzalez, Luis, 29

Goodstein, David L., 87

Görner, Hermann, 94

GPS, 204

Gravity

aging and, 204–206, 206 (table)

baseball traveling in "space stadium," 96–97

general relativity and, 204–206

golfing in solar system, 55–56, 56 (table)

jumping/altitude differences, 83–84

at surface of Earth, 55

Gravity equations

between objects, 90–91

at surface of Earth, 90

Green, A. C., 42

Grip strength, 94

Guidry, Ron, 32n13

Halladay, Roy, 24

Handcuffs problem, 197–198

Handey, Jack, 98

Hanks, Tom, 22

Harmonic motion

boobs, 130–132, 133

coupling and, 132–133

damping and, 132

description, 128–129

resonance and, 133

Heat capacity, 136

Heder, Jon, 124

Heiden, Eric, 24–25

Helium-filled football, 142–145, 145 (table)

Hertz, 123

Hertz, Heinrich, 123

Hockey players

goalie size/blocking goal, 40–41

teeth losses estimates, 14–15

teeth losses/money saved on toothpaste, 18–19

Hogan, Hulk, 46

Holyfield, Evander, 90

Hooke, Robert, 128

Hooke's law, 128, 128n83

Hot air balloons, 135

Hot dogs and rat droppings, 21

How Many Licks? Or, How to Estimate Damn Near Anything (Santos), 53n31, 210n138

Howard, Elston, 32n13

Ideal gas law, 135

Inertia, 88n53

International System of Units, 9n4

International Weightlifting Federation (IWF), 98

Invisibility vs. strength choice, 103

Jackson, Bo, 38

Jackson, Reggie, 32n13

James, LeBron, 46
Jerk (movement) defined, 62
Jeter, Derek, 32, 195, 195n130
Johnson, Chris, 76
Jordan, Michael, 25, 29, 57, 59, 89, 193, 196
Joule, 107n71
Joule, James Prescott, 107n71
Joyce, Jim, 199–201
Jumping
  altitude differences and, 83–84
  Mario the plumber, 38–39
  off Empire State Building, 140–141

Kasparov, Garry, 174
Kelvin/Celsius conversion, 135n89
Kinetic energy of motion, 107
Kinsey, Alfred, 166
Knee pressure, O'Neal vs. Bogues, 51–52
Kobayashi, Takeru, 109–110

Lackey, John, 24
Landreaux, Ken, 176
League of Their Own, A (movie), 22
Leagues now defunct, 31
Lesnar, Brock, 90
Life magazine, 87
Light speed, 69–70
Longoria, Evan, 142

Mad Lib, 60, 60n34, 61
Madden NFL video game series, 38
Magnus Effect/force, 149, 158
Magnus, Heinrich, 149
Man in the Middle (Amaechi), 166
Mantle, Mickey, 32n13, 195
Marathons/relativistic effects, 202–203
Marino, Dan, 10

Mario the plumber, 38–39
Maris, Roger, 32n13
Martin, Billy, 32n13
Martin, Demetri, 6
Mass
  definition, 51n29
  of Earth, 100
  weight vs., 51n29, 92n58
Mathematics popularity, 5–6
Matthews, Clay, 88
Mattingly, Don, 32n13
McHale, Kevin, 16
Mendoza, Mario, 25, 180–181, 180n119, 193, 196
Micron vs. millimeter, 35n18
Microsecond, 126n81
Millimeter vs. micron, 35n18
Momentum conservation
  description, 122
  Newton's laws and, 90n55
  See also Angular momentum
Monroe, Marilyn, 176n116
Monty Hall doors/problem, 188–189
Moon golfing, 55, 56 (table)
Motion equation, 63–64, 63n35
Motion of projectiles
  air friction and, 80
  considerations with, 78–79
  gravity and, 78–79, 85–87
  hitting never-landing ball, 85–87
  horizontal range equation, 79–80
  overview, 78–80
  punting/hitting blimp, 81–82
  subscripts in equations, 78n43
  vectors and, 78
Munson, Thurman, 32n13
Muscleless Wonder, 92, 112, 114, 118, 140, 142, 210
MythBusters (television), 94n61, 143, 145

National Lampoon's European Vacation (movie), 148n98

NBA Jam (video game), 146

"Newton" as unit of force, 91n57

Newton, Isaac/laws
overview, 88–91
using laws, 94, 95, 107n72, 149, 153, 157, 161

Niekro, Phil, 169

Nintendo Entertainment System (NES), 37, 96

Nixon, Richard, 137n91

Noninertial frames, 203n135

Norwood, Scott, 44

Olympics, 5, 24–25, 74, 83, 111, 137, 154

"Omega" symbol defined, 123

O'Neal, Shaquille, 51–52, 204–206

"Order of magnitude" defined, 7–8

"Order parameters," 46, 194, 194n129

Oscillation, 128–129
See also Harmonic motion

Parish, Robert, 32n14

"Perfect" games, 193–194

Perry, Douglas, 130

Phelps, Michael
Saturday Night Live, 111
swimmers/overflowing pool, 34–35
swimming with cement loafers, 154–156, 155 (table)

Physics Fact Book, The, 11, 150

Ping-pong
ball breaking sound barrier, 137–139
Beijing Olympics, 137
Chinese-US relations and, 137n91

Pippen, Scottie, 29

Pisarenko, Anatoly, 53–54

Playoffs
AL West/NL Central probabilities, 164–165
ticket sales/finances and, 165

Pole vaulting Gateway Arch, Saint Louis, 120–121

Power defined, 108

Prayers for winning, 27–28

Pressure
description/equation, 51, 134–135
of gas/ideal gas law, 135
ping-pong ball speed, 137–139

Price, Mark, 57–58

Probability
of 56-game hitting streak, 176–179, 180–181
baseball card packs/card set, 185–188, 191
baseball card packs/card set with trading, 191–192, 191n128
buzzer beaters, 57–59
catching home run in stands, 170–172, 171n114
conditional probability, 188–189
Crawford's ref record, 182–185, 184 (table)
of gay NFL Hall of Famer, 166–167, 167n111
Monty Hall doors/problem, 188–189
overview, 163
playoff advantage, 164–165
random coincidences, 168–169
Six Degrees games, 168–169, 168n112

Pryor, Richard, 20

Pujols, Albert, 108

Punting/hitting blimp, 81–82

Pythagorean theorem, 80n45

Radio (sports radio) illogical calls estimation, 24–26

Rat droppings in hot dogs, 21

Reaction time
baseball players/steroids, 104
calculating fastest time, 72–73
description, 71–72

"Reference frame," 200–201, 202–203

Religion and sports, 27–28

Resonance and harmonic motion, 133
Reynolds number, 149n100
"Rho" symbol defined, 93n59, 134
Rice, Jerry, 10
Rising fastball, 157–158, 158 (table)
Rizzuto, Phil, 32n13
Robinson, Jackie, 32n13, 69–70
Rocky movies, 92–93
Rodriguez, Alex, 7
Roenick, Jeremy, 38
Roethlisberger, Ben, 46
Romine, Kevin, 185
Rope handcuffs problem, 197–198
Running
    direction on track, 126–127, 126n80
    marathons/relativistic effects, 202–203
    outermost lane numbers/length, 50
    See also specific individuals
Running of the bulls, Spain, 8
Ruth, Babe, 32n13, 193, 195, 196

Sabathia, C. C., 176
Sabermetrics, 5, 6, 194, 196n132
Saturday Night Live, 111
Scaling arguments
    complex vs. simple systems, 57–59
    described, 49
    golfing in solar system, 55–56, 56 (table)
    jumping/altitude differences, 83–84
    See also specific examples
Scientific notation, 7–9, 9n3
Secretariat, 195
Selig, Bud, 164
Sensabaugh, Gerald, 39
Sex and weight loss, 118–119
Shepard, Alan, 55
"Sigma" symbol defined, 177n117
Simmons, Bill, 182
Simpsons, The, 101, 102

Sine of angle, 80n45
Six Degrees
    of Hank Aaron, 168–169
    of Kevin Bacon, 168
Skee ball game theory, 33
Slice, Kimbo, 46
Smith, Emmitt, 10
Sneakers' tread loss, 45
Sopranos (television), 154
Sound. See Speed of sound
South Park, 137
Spahn, Warren, 168
Special relativity, 200–201, 202–203
Speed
    Bolt vs. Johnson, 76–77
    Bolt vs. racecar, 74–75
    overview, 62–65
    See also specific examples
Speed of electricity, 69–70, 69n41
Speed of light, 69–70
Speed of sound
    estimating, 66–67
    ping-pong ball example and, 137–139
    starting pistol/runners and, 68
Spitting spectators, 44
Sports radio illogical calls estimation, 24–26
Stair climbing and weight loss, 112–113
Stallone, Sylvester, 92–93
STDs contracted/Chamberlain, 42–43
Stengel, Casey, 32n13
Steroids. See Baseball players/steroids
Stokes' Law, 148n98
Strength
    baseball players/steroids and, 103–106
    vs. invisibility choice, 103
Sumo wrestler's fat/stopping bullet, 159–162, 162 (table)

Swimming
with cement loafers, 154–156, 155 (table)
pool overflowing/number of swimmers, 34–35
Szuminski, Jason, 169

Tangent of angle, 80n45
Taylor, Brook, 63n35
Taylor, G.I., 52
Taylor series, 63n35, 128n83, 148n98, 205n137
Tecmo video games, 38, 130
Teeth losses. See Hockey players
Temperature
atom movement/change in energy, 136
jumping and, 140–141, 141n95
Kelvin/Celsius conversion, 135n89
Theory, "just a theory," 200n133
Theory of mind, 24, 24n10
Thomson, William, First Baron Kelvin, 135n89
Thunderstorm distance estimating, 66n37
Tipping
autograph vs., 29–30
good tippers/bad tippers, 29
Toews, Jonathan, 47
Topps baseball cards, 185–188
Torre, Joe, 32
Tour de France, 8, 24, 25
Trigonometry overview, 80n45
Tyson, Mike, 38, 90, 174
Uncertainty principle, 209
Units
converting, 10, 11
overview, 9–10, 9n4

Vectors and motion of projectiles, 78
Velazquez, John, 122

Velocity
definition, 62, 86n50
equation of motion and, 63–64, 63n35
Vick, Michael, 142
View from Above, A (Chamberlain), 42
Viscosity, 148

Wear and tear rate, 45
Webber, Chris, 44
Websites
to avoid, 11
for converting units, 11
Weight
definition, 51n29
mass vs., 51n29, 92n58
scaling with length, 53–54
See also Energy/weight loss
Weight lifting
by ants, 53–54, 54n32
bench presses/weight loss, 114–115
energy conversion and, 107–108
flatulence effects, 98
pulleys and, 99–100, 100n65
record for one-handed deadlift, 94
shrinking weight lifter, 53–54
by Stallone (Rocky), 92–93
urination and, 98
Wells, David, 169
White, Will, 176
Wikipedia, 11, 25, 37, 72, 94, 174
Wilfork, Vincent, 122
Williams, Kyle, 44
Williams, Ted, 71, 176
Woods, Tiger, 42
Work defined, 107n71

Young, Cy, 176

Zedong, Zhuang, 137n91
Zito, Barry, 103

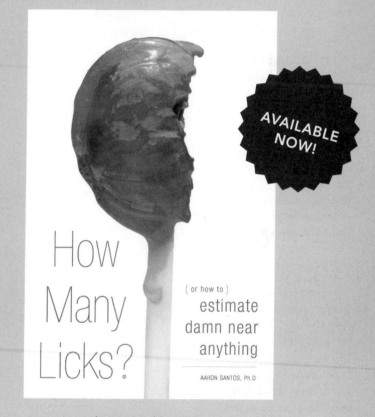

—